Fundamentals and Applications of Organic Electrochemistry

Fundamentals and Applications of Organic Electrochemistry

Synthesis, Materials, Devices

Toshio Fuchigami and Shinsuke Inagi
Department of Electronic Chemistry,
Tokyo Institute of Technology, Japan

Mahito Atobe
Department of Environment and System Sciences,
Yokohama National University, Japan

This edition first published 2015

An earlier version of this work was published in the Japanese language by Corona Publishing Co. Ltd under the title 有機電気化学 - 基礎から応用まで -

Registered office
John Wiley & Sons Ltd, The Atrium, Southern Gate, Chichester, West Sussex, PO19 8SQ, United Kingdom

For details of our global editorial offices, for customer services and for information about how to apply for permission to reuse the copyright material in this book please see our website at www.wiley.com.

Library of Congress Cataloging-in-Publication Data

Fuchigami, Toshio.
Fundamentals and applications of organic electrochemistry : synthesis, materials, devices / Toshio Fuchigami, Mahito Atobe, Shinsuke Inagi.
pages cm
Includes bibliographical references and index.
ISBN 978-1-118-65317-3 (cloth)
1. Organic electrochemistry. 2. Electrochemistry. 3. Chemistry, Organic. I. Atobe, Mahito, 1969- II.
Inagi, Shinsuke. III. Title.
QD273.F83 2014
547′.137–dc23
2014017644

A catalogue record for this book is available from the British Library.

ISBN: 9781118653173

Set in 10.5/13pt, SabonLTStd by Thomson Digital, Noida, India.

1 2015

Contents

About the Authors

Dr Toshio Fuchigami is an institute professor at the Tokyo Institute of Technology, having received his PhD from the same institute in 1974. He has authored more than 420 publications, including review articles and book chapters. His current research interests are centred on the new hybrid fields of organofluorine electrochemistry and new electrolytic systems for green sustainable chemistry. He has received many awards in electrochemistry, including the Electrochemical Society of Japan Award and the Electrochemical Society (ECS) Manuel M. Baizer Award. He is also an ECS fellow.

Dr Mahito Atobe was appointed to a professorship at the Graduate School of Environment and Information Sciences, Yokohama National University in July 2010. He received his PhD from the Tokyo Institute of Technology in 1998. His current research focuses on organic electrosynthetic processes and electrochemical polymerisation under ultrasonication, electrosynthetic processes in a flow microreactor, and organic electrochemical processes in supercritical fluids (140 publications).

Dr Shinsuke Inagi is an associate professor at the Tokyo Institute of Technology. He received his PhD from Kyoto University in 2007. After postdoctoral research (Research Fellowship for Young Scientists from the Japan Society for the Promotion of Scientists) at Kyoto University, he joined Professor Fuchigami's research group as an assistant professor at the Tokyo Institute of Technology in 2007. He was promoted to associate professor in 2011. His current research interests include electrochemical synthesis of polymeric materials.

Preface

Organic electrochemistry is electrochemistry focused on organic molecules, while inorganic electrochemistry deals with inorganic molecules, which is a major part of fundamentals and applications of electrochemistry. In fact, most industrialized electrode processes are inorganic. Electrochemistry is mainly based on physical chemistry such as thermodynamics and kinetics. Many mathematical equations are used in electrochemistry textbooks and organic chemists therefore think that electrochemistry is difficult. Similarly, organic chemistry deals with organic molecules and complicated reactions, therefore physical chemists often dislike organic chemistry.

Organic electrochemistry, particularly organic electrosynthesis, has developed by incorporating new organic reactions and organic synthesis. The 21st century is sometimes called the ecological century, and organic electrosynthesis is a typical green sustainable chemistry since it does not require any hazardous reagents and produces less waste than other chemical synthesis. Furthermore, organic electrochemistry has also recently developed as integrated field including not only organic electro-synthesis but also materials chemistry, catalysis chemistry, biochemistry, medicinal chemistry and environmental chemistry. In our daily lives organic and polymer materials play important roles in technologies such as biosensors, conducting polymers, liquid crystals, electroluminescence materials, dye-sensitized solar cells and so on. To understand these technologies we must study the basics of both organic chemistry and electrochemistry. In this century, the area of interest is diversity, therefore students, particularly graduate students, can no longer be engaged in developments in cutting-edge technology unless they understand the fundamental principles of various sciences such as organic chemistry, inorganic chemistry and physical chemistry, regardless of their own specialized scientific background.

In addition, organic electrochemistry also involves organic electron transfer chemistry using electrical energy. In this way organic electrochemistry is quite similar to photoelectron transfer, which is an important field of organic photochemistry using light energy. Although a number of fundamental books dealing with organic photochemistry have been published, there has been no textbook dealing with the basic aspects of organic electrode electron transfer and its applications together with new fields.

In this book, the authors have concisely produced their organic electrochemistry lecture notes for graduate students. The text is arranged for graduate students, researchers and engineers to easily understand the basic principles of electrochemistry, electrochemical measurements and organic electrosynthesis, including its new methodologies. Some experimental examples of organic electrosynthesis are also described in detail.

Online supplementary material for the book can be found at http://booksupport.wiley.com

Introduction

Toshio Fuchigami

The concept of organic electrochemistry is relatively new, even though it has a long history. In 1800, an Italian physicist named Volta invented the well-known Voltaic pile. Three years later, Petrov in Russia published a paper on the electrolysis of alcohols and aliphatic oils. A year after that, Grotthuss in Lithuania, who proposed the ionic conducting mechanism, found that a diluted solution of indigo white could be readily electrochemically oxidized to indigo blue. In 1833, Faraday discovered Faraday's law, and one year later he found that hydrocarbons could be formed by the electrolysis of an aqueous solution of the acetic acid salt. Unfortunately, he could not identify the products. In 1849, Wöhler's disciple, Kolbe, discovered the electrochemical oxidation of a carboxylic acid (RCOOH) to the dimeric alkane (R–R) and CO_2, known as Kolbe electrolysis [1]. Consequently, Faraday and Kolbe are pioneers in the investigation of organic electrochemical processes. From the end of the 19th century to the early 20th century, electrochemical oxidation and reduction processes of various compounds were intensively investigated. Thus, the application of electrolysis for preparing organic compounds continued in the first half of the 20th century. A typical example is the electrochemical reduction process of nitrobenzene to aniline. Importantly, organic electrochemistry was also developed along with the discovery of new electroanalytical techniques such as polarography, which was developed by Heyrovský and Tachi in the early 1920s [2]. However, organic electrosynthesis research had to be completely halted during the Second World War.

In 1964, Baizer developed the electrochemical hydrodimerization of acrylonitrile, which is a highly useful industrial process for the manufacture of adiponitrile. This invention restimulated organic electrosynthesis research by many electrochemists and organic chemists. Since then, the development of organic electrochemistry, particularly organic electrosynthesis, has been marked by incorporating new types of organic reactions and modern organic synthesis. Furthermore, various aprotic polar organic solvents have been developed, and these enable us to detect electrogenerated unstable intermediates. In addition, cyclic voltammetry and related electroanalytical techniques have assisted in the understanding of kinetics and mechanisms of organic electrode processes.

Organic electrochemistry has recently developed as an integrated field including not only organic electrosynthesis but also materials chemistry, catalysis chemistry, biochemistry, medicinal chemistry and environmental chemistry.

The 21st century is known as the ecological century. Organic electrosynthesis is expected to be a typical green chemistry process since it does not require any hazardous reagents and produces less waste than conventional chemical synthesis. Recently, a novel paired electrosynthesis of phthalide and *p*-*t*-butyl benzaldehyde has been developed and industrialized by BASF in Germany, and they consider electrosynthesis to be the most promising green synthetic process. These facts have prompted organic electrochemists as well as organic chemists to make great efforts to develop new systems of organic electrosynthesis in order to achieve green and sustainable chemistry. In fact, a number of successful new green organic electrolytic systems have been developed to date, as illustrated in this book. We believe that cutting-edge developments in organic electrochemistry will be achieved through hybridization with other scientific fields, as mentioned above.

REFERENCES

1. Vijih, A.K. and Conway, B.E. (1967) *Chem. Rev.*, **67**, 623–664.
2. Zuman, P. (2012) *Chem. Rec.*, **12**, 46–62.

1

Fundamental Principles of Organic Electrochemistry: Fundamental Aspects of Electrochemistry Dealing with Organic Molecules

Mahito Atobe

Chemists often encounter situations in which a reaction does not proceed at a convenient rate under the initially selected set of conditions. In chemistry, activation energy is defined as the minimum energy required to start a chemical reaction, and hence the activation energy must be put into a chemical system in order for a chemical reaction to occur. Catalysts are often used to reduce the activation energy but a high temperature is still required for the reaction to proceed at an appreciable rate. Electrochemical reactions, however, can generally be carried out under mild conditions (room temperature and ambient pressure).

Fundamentals and Applications of Organic Electrochemistry: Synthesis, Materials, Devices, First Edition. Toshio Fuchigami, Mahito Atobe and Shinsuke Inagi.

In electrochemical reactions there is an additional experimental parameter, the electrode potential, involved in the manipulation of electrochemical reaction rates. Electron transfer rates can easily be varied over many orders of magnitude at a single temperature by proper control of the electrode potential. Indeed, electrode potential is so powerful a parameter for controlling the rates of electrochemical reactions that most reactions can be carried out at or near room temperature.

An understanding of the nature of the dependence of electron transfer rates on potential is important for understanding electrode processes and constitutes the central theme of this chapter. Because electron transfer at an electrode surface is necessarily a heterogeneous process, it will be necessary to examine briefly the structure of the electrode–solution interface and its effects on the course of an electrochemical reaction. It is not enough, however, to simply derive the relationship between electron transfer rate and electrode potential. This is because as a result of the dramatic changes in these rates with potential, it is generally found that at certain potentials electron transfer is so fast that the overall process is actually limited by the rate of mass transport of the substrate from the bulk solution to the electrode surface. There are different modes of mass transport, and they differ in efficiency, therefore it will be necessary to examine each of these influences.

1.1 FORMATION OF ELECTRICAL DOUBLE LAYER

When electrodes are polarized in an electrolyte solution, the charge held at the electrodes is important. In order to neutralize a charge imbalance across the electrode–solution interface, the rearrangement of charged species like ions in the solution near the electrode surface will occur within a few hundredths of a second, and finally result in strong interactions occurring between the ions in solution and the electrode surface. This gives rise to the electrical double layer, whose thickness is usually between 1 and 10 nm (Figure 1.1) [1]. There exists a potential gradient over the electrical double layer and the gradient is no longer confirmed to the bulk electrolyte solution. The potential difference between the electrode surface and the bulk solution illustrated in Figure 1.1 may amount to a volt or more, over the rather short distance of the thickness of the double layer, and hence this is an extremely steep gradient, in the order of $10^6\ V\ cm^{-1}$ or greater, which is an electrical field of considerable intensity. This is the driving force for the electrochemical reaction at electrode

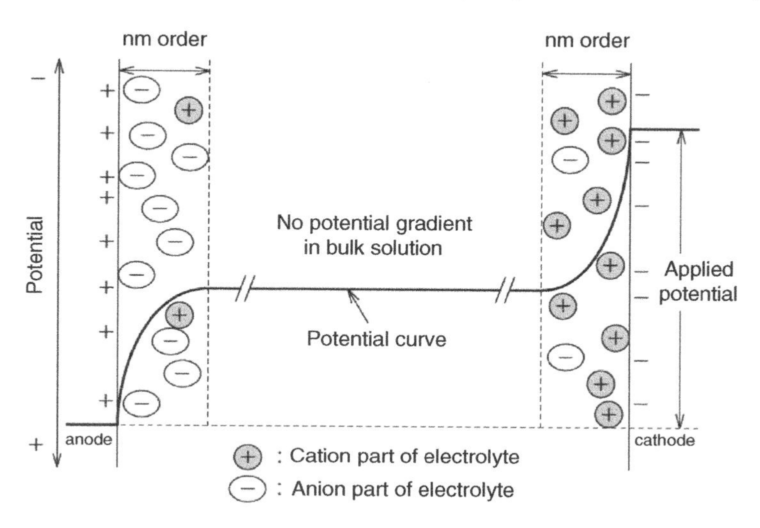

Figure 1.1 Electrical double-layer model and potential distribution in the double layer

interfaces, therefore when the polarization between anode and cathode is increased gradually, the potential gradient in the vicinity of the anode and cathode is also increased and consequently the most oxidizable and reducible species in the system are subject to an electron-transfer reaction at the anode and cathode, respectively. Because a charge imbalance in the vicinity of an electrode takes place after the electron-transfer reaction, ions are transferred to the electrode interface to neutralize the imbalance, and consequently the continued Faradic current is observed. Thus, the electrolyte in a solution plays a role in the formation of the electrical double layer and the neutralization of a charge imbalance after electrolysis.

1.2 ELECTRODE POTENTIALS (REDOX POTENTIALS)

In all electrochemical experiments the reactions of interest occur at the surface of the working electrode therefore we are interested in controlling the potential drop across the interface between the surface of the working electrode and the solution. However, it is impossible to control or measure this interfacial potential without placing another electrode in

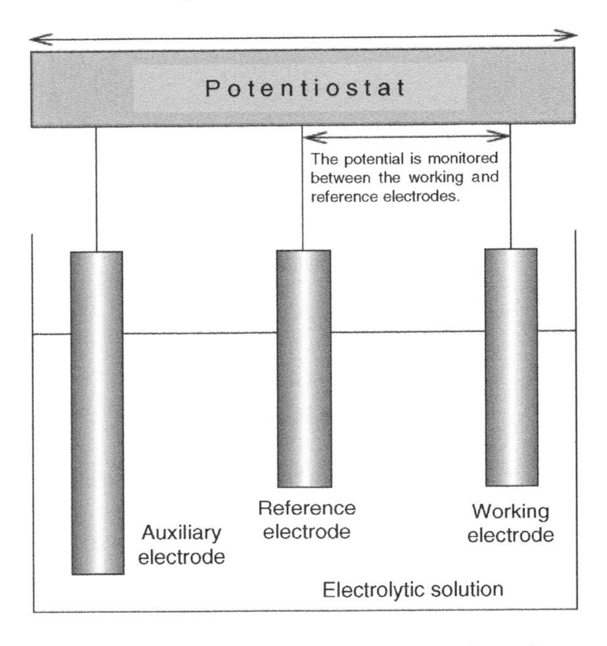

Figure 1.2 Experimental setup for the three-electrode system

the solution. Thus, two interfacial potentials must be considered, neither of which can be measured independently. Hence, one requirement for the counter electrode is that its interfacial potential remains constant so that any changes in the cell voltage produce identical changes in the working electrode interfacial potential. An electrode whose potential does not vary with the current is referred to as an ideal non-polarizable electrode, but there is no electrode that behaves in this way. Consequently, the interfacial potential of the counter electrode in the two-electrode system discussed above varies as the current is passed through the cell. This problem is overcome by using a three-electrode system in which the functions of the counter electrode are divided between the reference and auxiliary electrodes (Figure 1.2) [2]. This ensures that the potential between the working and reference electrodes is controlled and the current passes between the working and auxiliary electrodes. The current passing through the reference electrode is further diminished by using a high-input impedance operational amplifier for the reference electrode input.

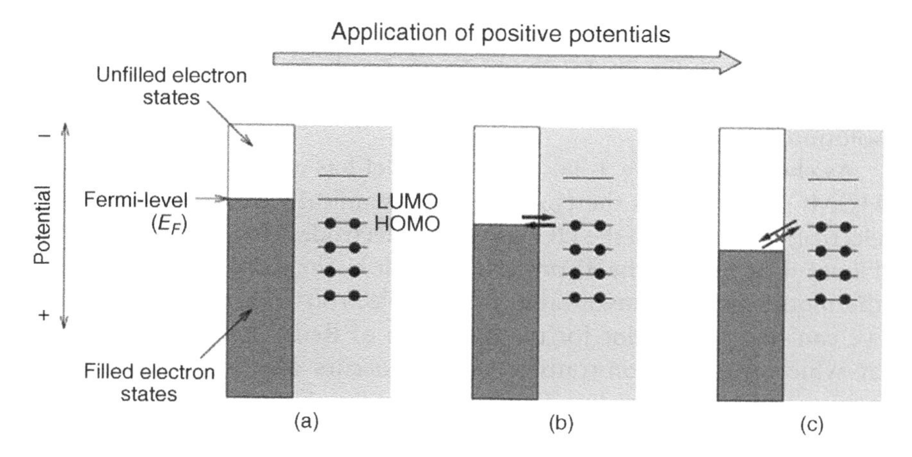

Figure 1.3 Fermi-level within a metal along with the orbital energies (HOMO and LUMO) of a molecule (Red) in solution

By employing the three-electrode system we can control or measure the working electrode potential. We then consider the essential meaning of the potential control using the following simple redox couple (Eq. 1.1):

$$\text{Red} \rightleftarrows \text{Ox} + ne^- \tag{1.1}$$

where Red and Ox represent the reduced and oxidized forms of a given species.

It is helpful to focus on the energy of electrons in the working metal electrode and in the Red species in the electrolyte solution, as depicted in Figure 1.3. The behaviour of electrons in a metal electrode can be partly understood by considering the Fermi-level (E_F) [3]. Metals are composed of closely packed atoms that have strong overlap between one another. A piece of metal therefore does not possess the individual well-defined electron energy levels that would be found in a single atom of the same material. Instead a continuum of levels exists, with the available electrons filling the states from the bottom upwards. The Fermi-level corresponds to the energy of the highest occupied orbitals (HOMO). This level is not fixed and can be moved by supplying electrical energy (see Figure 1.3). We are therefore able to alter the energy of the Fermi-level by applying a potential to an electrode (when a negative potential is applied, the Fermi-level moves to higher energy; when a positive potential is applied, it moves to a lower energy.). Depending on the position of the

Fermi-level it may be thermodynamically feasible to reduce/oxidize species in solution. Figure 1.3 shows the Fermi-level within a metal along with the orbital energies (HOMO and LUMO) of a molecule (Red) in solution.

As shown in Figure 1.3a, the Fermi-level has a higher value than the HOMO of Red. It is therefore thermodynamically unfavourable for an electron to jump from the HOMO to the electrode. However, as shown in Figure 1.3c, when the Fermi-level is below the HOMO of Red it is thermodynamically favourable for the electron transfer to occur and we can observe current for the oxidation of Red. The critical potential at which this electron-transfer process occurs identifies the standard potential, E°, of the redox couple Red/Ox (see Figure 1.3b).

1.3 ACTIVATION ENERGY AND OVERPOTENTIAL

As mentioned in section 1.2, depending on the relative position of the Fermi-level to the orbital energies (HOMO and LUMO) of a substrate molecule in solution, it may be thermodynamically feasible to reduce/oxidize the molecule. However, in general electrochemical reactions possess energy barriers that must be overcome by the reacting species. This energy barrier is called the activation energy (see Figure 1.4). Hence, the potential difference above the equilibrium value (the standard potential, E°) is usually required to produce a current. This potential difference

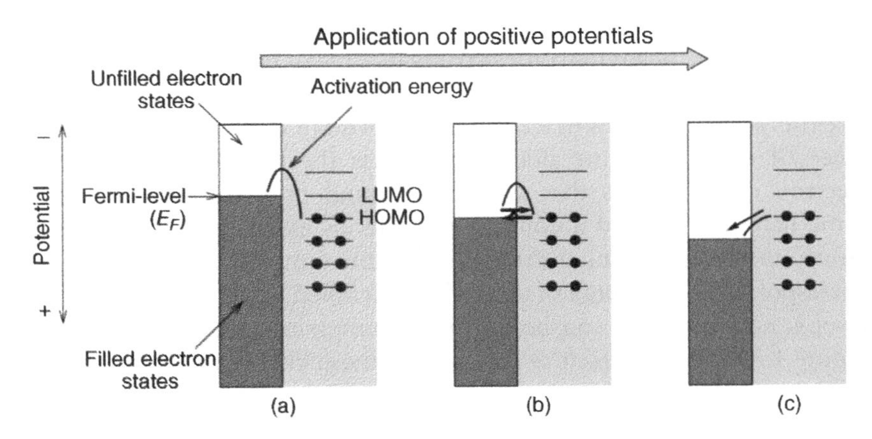

Figure 1.4 The activation energy in the electron transfer process at an electrode

between the standard potential and the potential at which the redox event is experimentally observed is called the overpotential [4].

1.4 CURRENTS CONTROLLED BY ELECTRON TRANSFER AND MASS TRANSPORT

Although electrode potential is an extremely important experimental parameter for manipulation of electrochemical reaction rates, other parameters such as mass transport can also affect reaction rates. We now consider the simplest electrochemical model, which is composed of the electron transfer process and the mass transfer process, as shown in Figure 1.5. In this case, depending on the electrode potential, the rate-determining step might be either the electron transfer rate or the rate of mass transport of the substrate to the electrode surface. To examine the quantitative and semiquantitative interrelationships between potential, electrochemical reaction rates and mass transport, a wide variety of voltammetric experiments are commonly used [5].

We will now consider the factors affecting the relative heights and shapes of voltammetric waves such as those in Figure 1.6. As the electrode potential is scanned during a voltammogram (to more negative potentials for a reduction or more positive potentials for an oxidation), the electron transfer rates are dramatically increased and a voltammetric curve passes through a mixed region in which the rates of mass transport and electron

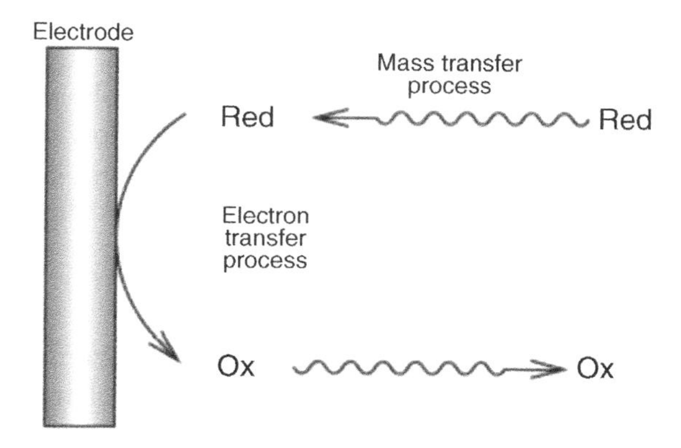

Figure 1.5 Electrochemical model showing the electron transfer process and the mass transfer process

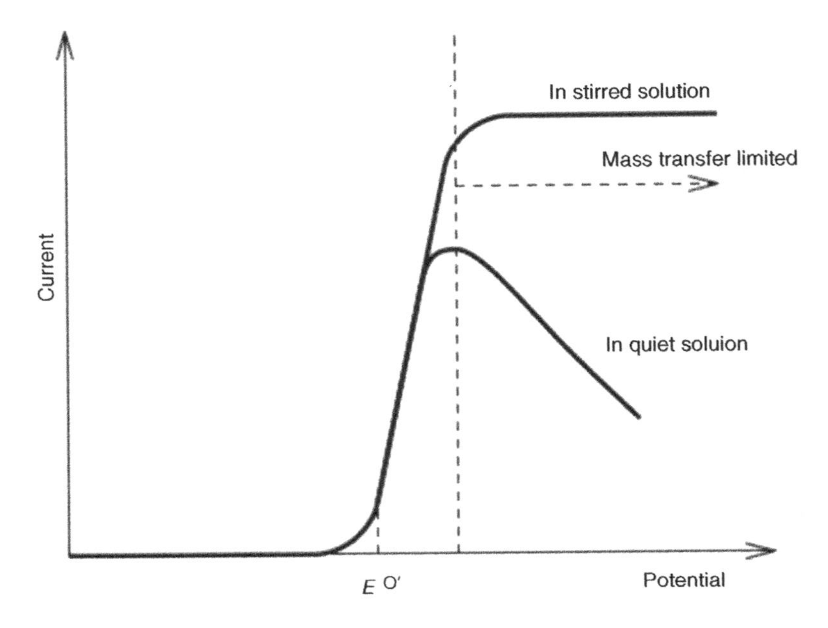

Figure 1.6 Voltammetric waves in quiet and stirred solution

transfer both limit the electrolysis current. Finally, the electron transfer rate eventually reaches a high enough rate that the currents are purely mass transfer limited, and the mass-transport-limited voltammetric peak and plateau currents are observed in quiet and stirred solution, respectively. For this reason the equations dealing with the electrolysis current are different for each rate-determining step.

When the electrochemical reaction is controlled by the electron transfer step, the net electrolysis current density (i) is represented by the Butler–Volmer equation (Eq. 1.2) [4]. This equation describes how the current density (i) on an electrode depends on the overpotential (η), considering that both a cathodic and an anodic reaction occur on the same electrode:

$$i = i_a - i_c = i_0\left[\exp(\alpha nF\eta/RT) - \exp\{-(1-\alpha)nF\eta/RT\}\right] \quad (1.2)$$

where i_a and i_c are the individual anodic and cathodic current densities, respectively, i_0 is the exchange current density, α is the charge transfer coefficient (its value lies between 0 and 1, frequently being about 0.5 at lower overpotentials), n is the number of electrons involved in the electrode reaction, F is the Faraday constant and η is the overpotential.

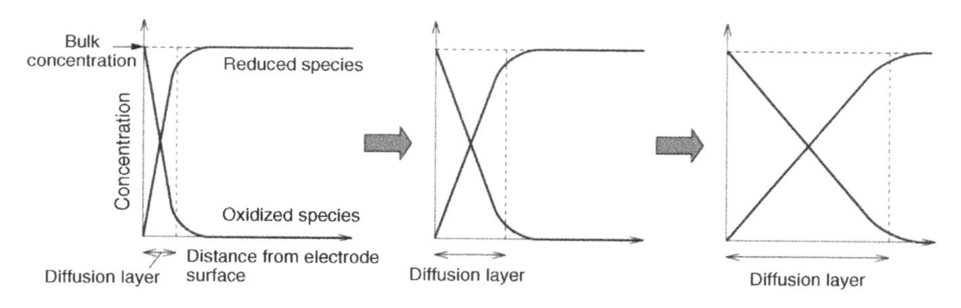

Figure 1.7 Change in the diffusion layer thickness with electrolysis time

As indicated by the Butler–Volmer equation, the net current density is the difference between the cathodic and anodic current densities. In addition, the exchange current density is that current in the absence of net electrolysis and at zero overpotential.

When the electrochemical reaction is controlled by the mass transfer step, the electrolysis current density relates to the magnitude of the gradient of the substrate molecule concentration at the electrode surface, represented by Eq. 1.3 [6]:

$$i = nFD(dc/dx)_{x=0} \tag{1.3}$$

where n is the number of electrons involved in the electrode reaction, F is the Faraday constant, D is the diffusion coefficient (the rate constant for motion of the substance through the given medium by diffusion), c is the substrate molecule concentration, x is the distance from the electrode surface and hence $(dc/dx)_{x=0}$ represents the gradient of the substrate molecule concentration at the electrode surface. The concentration profiles for the substrate and product in quiet solution are shown in Figure 1.7. Since the thickness of the diffusion layer and hence the concentration gradient change as electrolysis proceeds in quiet solution, the decrease in electrolysis current is observed in its voltammogram, as illustrated in Figure 1.6a.

In the presence of convection, for example stirring, the variation of the diffusion layer with time is inhibited and hence the concentration gradient is constant. In this case, a limiting current density (i_d) related to the diffusion layer thickness (δ) would be observed [5]:

$$i_d = nFDc/\delta \tag{1.4}$$

Equation 1.4 describes the current density at the plateau of a voltammogram measured in stirred solution (Figure 1.6b), provided that the potential is scanned rapidly, so the concentration of the substrate in the bulk of the solution is not significantly depleted during the time needed to measure the voltammogram. In addition, since δ is reduced by more efficient stirring, the limiting current densities in stirred solution increase with the stirring rate. They are of course much larger than the current densities in quiet solution because convection is so much more efficient than diffusion at transporting the substrate molecule to the electrode surface.

REFERENCES

1. Bard, A.J. and Faulkner, L.R. (2001) *Electrochemical Methods, Fundamentals and Applications*, 2nd edn, John Wiley & Sons, Inc., New York, Chapter 13.
2. Bard, A.J. and Faulkner, L.R. (2001) *Electrochemical Methods, Fundamentals and Applications*, 2nd edn, John Wiley & Sons, Inc., New York, Chapter 2.
3. Compton, R.G. and Sanders, G.H.W. (1996) *Electrode Potentials*, Oxford University Press, Oxford, Chapter 1.
4. Izutsu, K. (2009) *Electrochemistry in Nonaqueous Solutions*, Wiley-VCH Verlag GmbH, Weinheim, Chapter 5.
5. Fry, A.J. (1989) *Synthetic Organic Electrochemistry*, John Wiley & Sons, New York, Chapter 2.
6. Rifi, M.R. and Covitz, F.H. (1974) *Introduction to Organic Electrochemistry*, Marcel Dekker, New York, Chapter 2.

2

Method for Study of Organic Electrochemistry: Electrochemical Measurements of Organic Molecules

Mahito Atobe

One of the main aims of this book is to understand the mechanism of organic electrochemical reactions in order to use them most effectively or in new ways. Nowadays a number of specific electrochemical measurements are used to obtain mechanistic information about organic electrochemical reactions. Among these measurements, voltammetry is one of the most frequently used techniques since it provides enough evidence concerning the mechanism of an electrode process to allow us to use this process intelligently in synthesis and to develop new electrochemical reactions.

In this chapter we will introduce voltammetric techniques to obtain mechanistic information for organic electrochemical reactions. Because the success or failure of voltammetric measurements depends to a great

Fundamentals and Applications of Organic Electrochemistry: Synthesis, Materials, Devices, First Edition. Toshio Fuchigami, Mahito Atobe and Shinsuke Inagi.

degree on the proper selection of experimental components such as electrochemical cells, electrodes, solvents and electrolytes, we will discuss some concerns about experimental components for voltammetry. In addition, this chapter will also show how voltammetry can be used to obtain information about the mechanism of a new organic electrode reaction.

2.1 WORKING ELECTRODES

Voltammetry is the group of electrochemical techniques where current is studied as a response to the potential of the working electrode. Experiments are usually carried out using a three-electrode system in which the potential between the working and reference electrodes is controlled and the current passes between the working and auxiliary electrodes [1]. Since the measured electrochemical reaction occurs at the working electrode, the selection of the working electrode material is critical to the experimental success of voltammetry.

The electrolytes can be used without appreciable degradation only in limit ranges of electrical potential. This potential window should be as wide as possible to allow for the greatest degree of studied sample characterization [2]. The upper and lower potential limits are determined not only by the electrolytic solution but also by the electrode material. In aqueous electrolytes, oxygen and hydrogen evolution reactions limit the potential window, and hence the window is usually narrower than that in non-aqueous electrolytes. To overcome this problem, cathode materials with high hydrogen overpotential and anode materials with high oxygen overpotential are usually used as working electrode materials for voltammetry in aqueous electrolytes. Platinum and gold are good anode materials because of their higher oxygen overpotential, while mercury, zinc and lead are good candidates for cathode materials because of their higher hydrogen overpotential. However, the toxicity of mercury has lead to a limited use. On the other hand, the non-aqueous electrolytes are stable and their potential windows are generally larger than those of aqueous electrolytes, therefore there are few limitations for the choice of working electrode materials. The most commonly used working electrode materials for voltammetry in non-aqueous electrolytes are platinum, gold and carbon. Although almost all noble metals can be used as both anode and cathode materials, base metals are unsuitable as anode materials because of their dissolution under anodic polarization.

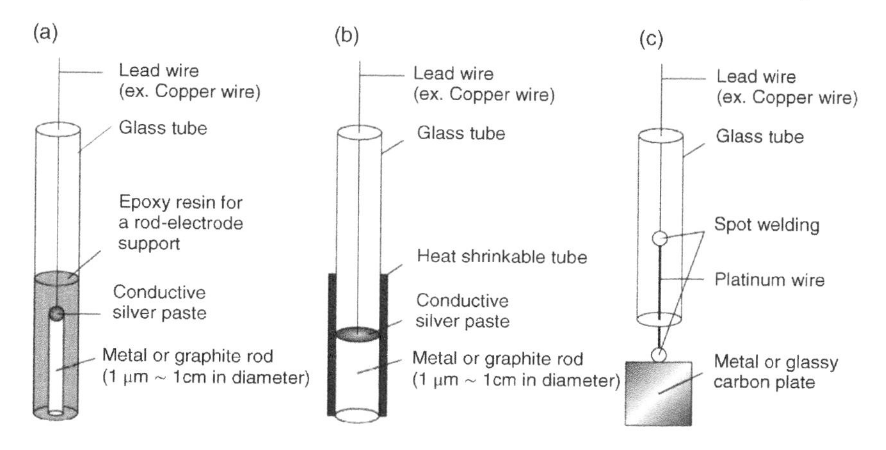

Figure 2.1 Examples of handmade working electrodes

Various shapes (e.g. disks, wires, plates) and sizes (e.g. a few square centimetres for plate electrodes, a few micrometres to a few centimetres in disk diameter) of solid electrodes are used for voltammetric measurements (Figure 2.1). Solid electrodes for voltammetric measurements are most often fabricated by encapsulating the electrode material in a non-conducting sheath of glass or inert polymeric material like Teflon, Kel-F (poly-chlorotrifluoroethylene) or PEEK (poly-etheretherketone). Most commonly, the exposed electrode material is in the form of a disk (Figures 2.1a and b). Common commercially available disk diameters range from 1 μm to 1 cm. Metal plate electrodes are usually connected to a lead wire by spot welding, and the wire part is encapsulated in a non-conducting sheath of glass or inert polymeric material (Figure 2.1c).

Ideally, a working electrode should behave in the same way each time it is used. The factors that affect the electrochemical behaviour of a surface are its cleanliness, the kind and extent of chemical functionalities (including oxides) that are present, and the microstructure of the electrode material itself. Generally, a pre-treatment step or steps will be carried out prior to each experiment to ensure that the electrode surface condition can be reproduced from experiment to experiment. These steps may be as simple as mechanical polishing, and may include pre-scanning across a certain potential range or exposure to a solvent or chemical species to activate the electrode. Specific procedures for different electrode materials can be found in references [1] and [2], and those contained therein.

2.2 REFERENCE ELECTRODES

The potential of the reference electrode must be stable and reproducible. In aqueous solutions, the method for measuring electrode potentials has been well established [3]. The standard hydrogen electrode (SHE) is the primary reference electrode and its potential is defined as zero at all temperatures. Practical measurements employ reference electrodes that are easy to use, the most popular ones being a saturated calomel electrode (SCE) (Figure 2.2a) and a silver–silver chloride (Ag/AgCl) electrode (Figure 2.2b). In contrast, in non-aqueous solutions the method for measuring electrode potential has not been established. The most serious problem is the reference electrode, that is, there is no primary reference electrode such as the SHE for non-aqueous electrolytes and no reference electrode as reliable as the aqueous Ag/AgCl electrode. However, efforts are being made to improve this situation.

The reference electrodes used in non-aqueous systems can be classified into two types. One type is an aqueous reference electrode, usually an aqueous Ag/AgCl electrode or SCE. However, the aqueous reference electrode should not be dipped directly into the non-aqueous solution under study because the solution is contaminated with water and the electrolyte (usually KCl). To prevent this, the reference electrode should be in a separate compartment and a salt bridge used for the ion-conducting connection between the working electrode and reference electrode

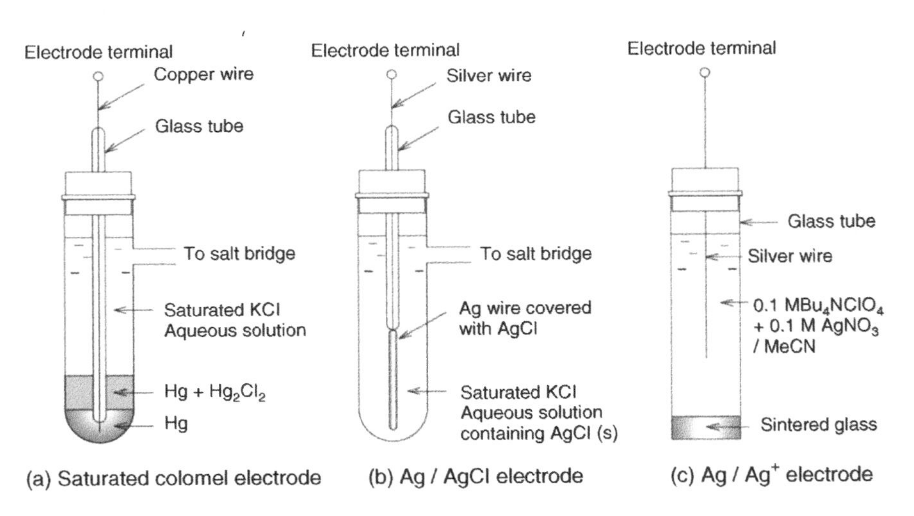

Figure 2.2 Examples of handmade reference electrodes

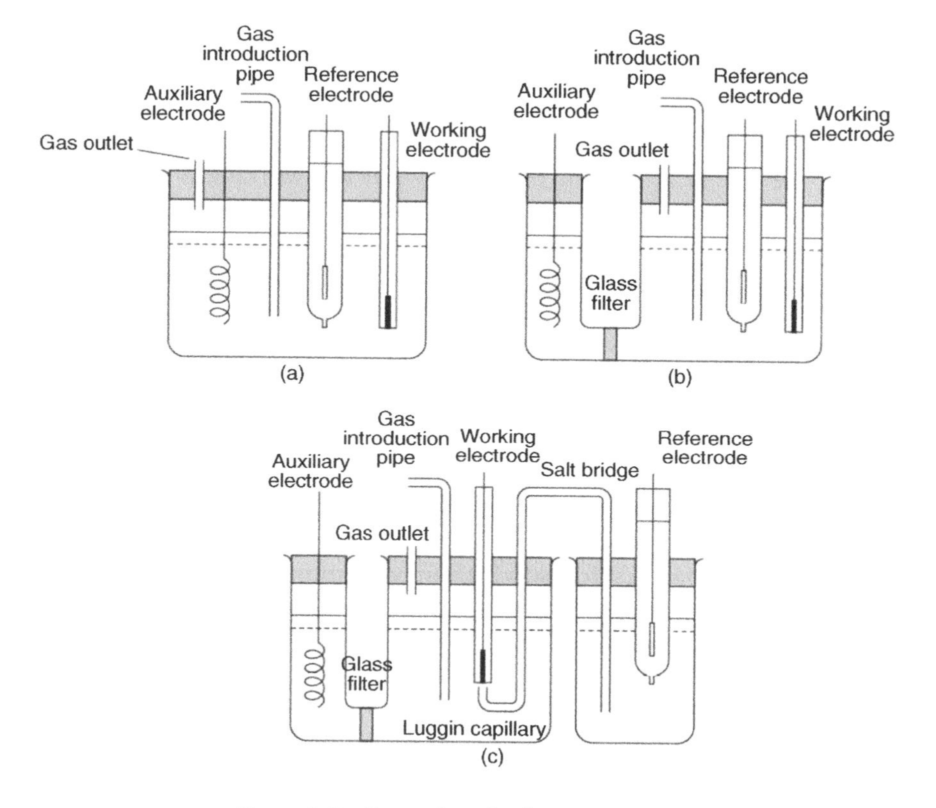

Figure 2.3 Examples of voltammetric cells

compartments (see Figure 2.3c). The tip of the salt bridge, which is filled with the non-aqueous electrolytes under study, is dipped into the non-aqueous solution. When such aqueous reference electrodes are used, the liquid junction potential between the aqueous and non-aqueous solutions must be taken into account. To improve this situation, the IUPAC Commission on Electrochemistry proposed that the Fc^{+}/Fc couple should be measured in the same system and the electrode potential reported as values referred to the apparent standard potential of the system. As for the other method, the same solvent as that of the solution under study is used for an internal solvent in the reference electrode. The Ag/Ag^{+} electrode is the most popular reference electrode used in non-aqueous solutions, and it can be used in a variety of solvents (Figure 2.2c).

Although the reference electrodes introduced in this section are commercially available, they can also be made in the laboratory.

2.3 AUXILIARY ELECTRODES

The auxiliary electrode, often also called the counter electrode, is an electrode used in a three-electrode electrochemical cell for voltammetry in which the electrical current passes between the working and auxiliary electrodes. The auxiliary electrode therefore functions as a cathode whenever the working electrode is operating as an anode and vice versa. The auxiliary electrode often has a surface area much larger than that of the working electrode to ensure that the half-reaction occurring at the auxiliary electrode can occur fast enough so as not to limit the process at the working electrode [4]. Platinum (e.g. Pt wire, Pt plate) is a good material for the auxiliary electrode due to its high stability.

2.4 SOLVENTS AND SUPPORTING ELECTROLYTES

The important factors for solvents in voltammetry are the potential window, the solubility of the substrate molecule under study and physical-chemical properties such as donor or solvating properties [5]. Although water is often used as a solvent for voltammetry, many organic substrates are water-insoluble or only sparingly soluble. In addition, the oxygen and hydrogen evolution reactions limit the potential window when water is used as a solvent for voltammetry. The voltammetric experiments for organic substrates are therefore usually carried out in polar organic solvents in which the supporting electrolyte can dissociate into ions. Acetonitrile (dielectric constant $\varepsilon = 38$) is one of the most frequently used solvents since its high upper and lower potential limits allow it to be used as a solvent for both electrochemical oxidation and reduction reactions. Other frequently used solvents for electrochemical oxidation are dichloromethane ($\varepsilon = 9$), nitromethane ($\varepsilon = 37$), propylene carbonate ($\varepsilon = 64$) and 1,2-dimethoxyethane ($\varepsilon = 3$), while other frequently used solvents for electrochemical reduction are *N*,*N*-dimethylformamide (DMF, $\varepsilon = 37$), dimethylsulfoxide (DMSO, $\varepsilon = 47$), tetrahydro-furan (THF, $\varepsilon = 7$) and benzonitrile ($\varepsilon = 26$). Although hexamethylphosphoramide (HMPA, $\varepsilon = 30$) is also a frequently used solvent for electrochemical reduction, extreme care must be used in handling it because of its toxicity.

A supporting electrolyte for use in voltammetry should fulfil the following conditions: (i) it should be soluble in the solvent under study and should dissociate into ions to give enough conductivity to the solution, (ii) it should be resistant to oxidation and reduction, and should give a wide potential window, and (iii) it should not have an unfavourable effect on the electrode reaction to be measured. In addition, the interaction between dissociated ions and intermediates formed by electrode reactions must be taken into account. For organic solvents, the commonly used electrolytes are tetraalkylammonium salts. In general, tetraethylammonim ion (Et_4N^+) and tetrabutylammonium ion (Bu_4N^+) are frequently used as the cation part of tetraalkylammonium salts, while perchlorate ion (ClO_4^-), tosylate ion (TsO^-), tetrafluoroborate ion (BF_4^-) and hexafluorophosphate ion (PF_6^-) are frequently used as the anion part. Because halide ion (X^-) may be oxidized to form halonium ion (X^+), it is necessary to be careful when using it as the anion part of the supporting electrolytes. On the other hand, for aqueous systems, inorganic salts such as NaCl and KCl, although not tetraalkylammonium salts, can be employed as supporting electrolytes. In practice, the concentration of supporting electrolytes should preferably be above 0.1 M.

2.5 CELLS AND POWER SOURCES

The most important decision to be made in planning voltammetry is whether to use an undivided cell, in which the working, auxiliary and reference electrodes are immersed in a single chamber (Figure 2.3a), or a divided cell, in which the working and auxiliary electrodes are in separate compartments (Figure 2.3b) [6]. In making this decision, we have to consider interference by electrolysis products at the auxiliary electrode in voltammetric experiments. If this happens a divided cell must be used. When the aqueous reference electrode, such as an Ag/AgCl electrode or SCE, is used in non-aqueous systems the reference electrode is also in a separate compartment in order to prevent contamination with water (Figure 2.3c), as mentioned in section 2.2. In this case, a salt bridge is used for the ion-conducting connection between the working electrode and reference electrode compartments, and the tip of the salt bridge is dipped into the non-aqueous solution. In addition, the tip should be as close as possible to the working electrode to minimize the iR drop between the tip and the electrode.

A potentiostat is used as a power supply and is fundamental to voltammetry using three-electrode systems [6]. In addition, a wave generator is necessary for the non-steady state measurements like cyclic voltammetry (CV). Instrumentation for CV, consisting of both the wave generator and potentiostat in a single package, is available from a number of manufactures. The voltammograms obtained are output to an X–Y recorder or PC.

2.6 STEADY-STATE AND NON-STEADY-STATES POLARIZATION CURVES

A plot of current density against electrode potential under a set of constant operating conditions, known as a polarization curve, is the standard electrochemical technique for characterizing the electrode reactions [7]. A steady-state polarization curve describes the relationship between the electrode potential and the current density, which is recorded by holding the electrode potential and recording the stable current response. Under vigorous stirring, the stable current response can be also obtained by potential scanning measurements and the diffusion limiting current is dependent on the potential scan rate. The voltammogram obtained in this case corresponds to the steady-state polarization curve. A non-steady-state polarization curve can be obtained at a rapid potential scan rate in the absence of any convections. However, even under still conditions a steady-state polarization curve may be obtained by using a slow potential scan since the influence of thermal convection cannot be ignored in this case.

To evaluate the limiting current density, a steady-state polarization curve is usually measured and the diffusion coefficient (D), diffusion layer thickness (δ) or the number of electrons involved in the electrode reaction (n) can be derived from the measured current density and Eq. 1.4.

In order to obtain mechanistic information about organic electrochemical reactions, non-steady-state polarization curves such as cyclic and linear sweep voltammograms are usually measured. CV is the most frequently used technique in organic electrochemistry, and hence we will discuss it in more detail below.

In CV, the potential of a stationary electrode is changed linearly towards increasingly negative or positive directions until the electrode process of interest, either a reduction or oxidation, respectively, takes place, and then the direction of the potential scan is reversed. This experiment is capable of providing a great deal of useful information about the redox potential of the studied organic molecules and the relative rates of electron transfer, mass transport and any chemical reactions taking place at the electrode surface.

Let us now consider the very simple CV for the redox reaction of ferrocene (Fc). It is well known that the redox reaction of ferrocene is represented by Eq. 2.1.

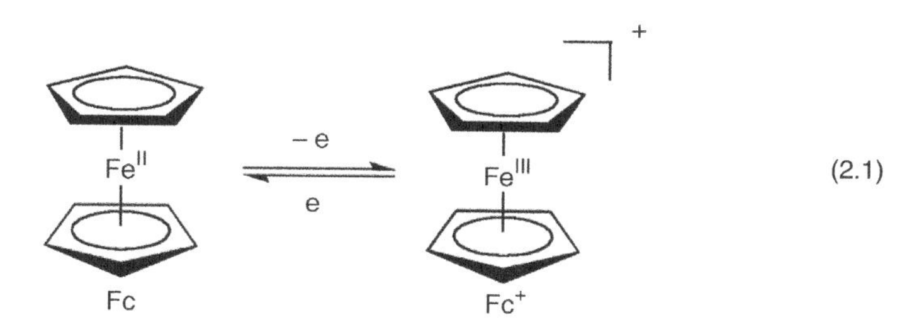

Because the Fc is the reduction state in this redox reaction, the potential should be swept firstly towards the positive direction (anodic scan). In this case, the potential scan should begin from the initial potential where no current flows (rest potential). In addition, the scan rate is generally set in the range between 50 and 200 mV s^{-1}. As mentioned in section 2.2, by changing the potential to an electrode the energy of the Fermi-level is also altered intentionally. When the potential is swept towards the positive direction from the rest potential (process a–b of Figure 2.4; the Fermi-level moves to a lower energy), the rate of electron transfer from the HOMO of

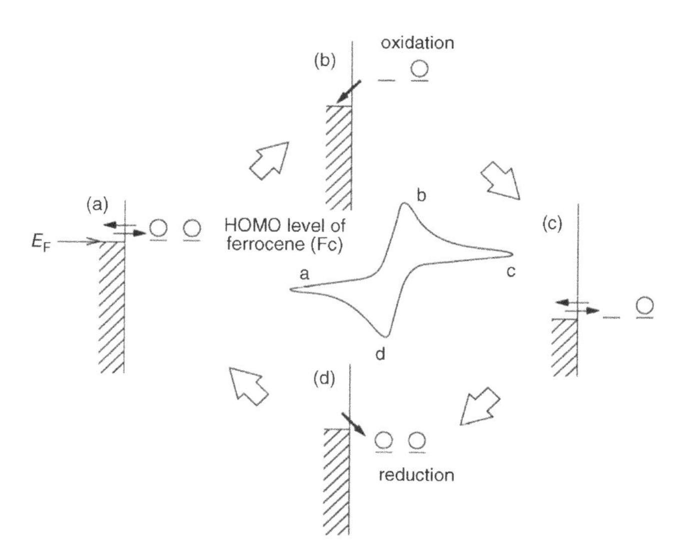

Figure 2.4 Cyclic voltammogram of ferrocene (Fc) and Fermi-level (E_F) within a metal electrode along with the HOMO level of Fc in solution

ferrocene to the Fermi-level of the electrode is dramatically increased and the potential passes through a mixed region in which the rates of mass transport and electron transfer both limit the current. When the electron transfer rate is high enough the currents are purely mass transfer limited, and a mass-transport-limited voltammetric peak is observed (Figure 2.4b). When the final potential is reached (Figure 2.4c), the direction of the potential scan is reversed (process c–d of Figure 2.4). Since the Fermi-level moves to a higher energy by doing this reverse scan, electron transfer from the Fermi-level to the HOMO of the ferrocenium ion (Fc^+) should occur (the current for Fc^+ reduction is observed in this case.). When most of the Fc^+ in the vicinity of the electrode surface has been converted to Fc, the reduction current is reduced (process d–a of Figure 2.4).

The form of cyclic voltammogram for ferrocene redox represents a typical reversible process. The oxidation peak current (i_p^a) in a reversible process like a ferrocene redox reaction is:

$$i_p^a = 0.4463nFc_{Fc}D_{Fc}^{1/2}(nFv/RT)^{1/2} \tag{2.2}$$

where n is the number of electrons involved in the electrode reaction (1 in the case of the ferrocene redox reaction), c_{Fc} and D_{Fc} are the bulk concentration and the diffusion coefficient of ferrocene, respectively, and v is the potential scan rate. Thus for a reversible CV wave, the peak current is proportional to the bulk concentration of a substrate or the square root of the potential scan rate.

However, the reversible redox reaction is uncommon in organic electrochemistry since many organic electrochemical reactions involve a fast chemical reaction subsequent to electron transfer (an EC process: the electrode step is E and the chemical step is C). In this case, the re-oxidation or re-reduction peak may totally disappear in the reversal potential scan.

2.7 POTENTIALS IN ELECTROCHEMICAL MEASUREMENTS

In the articles and books dealing with organic electrode reactions, there are many kinds of potentials [8,9]. The following six kinds of potentials are frequently used and hence we will discuss them:

1. standard electrode potential
2. formal potential

3. peak potential
4. half-peak potential
5. half-wave potential
6. decomposition potential (onset potential).

If we know the standard state free energy change, ΔG^o, for a chemical process, we can calculate the standard electrode potential, E^o, for an electrochemical reaction based on that process using the relationship between ΔG^o and E^o. Thus the E^o is not the experimental value but the calculated value. Therefore, even if the electrode potential is measured using the SHE under an equivalent state, the value measured will deviate from the real E^o value to a small degree. The value measured using the SHE under an equivalent state is termed the formal potential ($E^{o'}$). The relationship between E^o and $E^{o'}$ is represented by Eq. 2.3:

$$E^{o'} = E^o + \frac{RT}{nF}\ln\frac{\gamma_O}{\gamma_R} \tag{2.3}$$

where γ_O and γ_R are the activity coefficients for the oxidized and reduced forms of a studied species. However, the deviation between E^o and $E^{o'}$ is very small in a dilute solution and is usually less than 300 mV even in a concentrated solution.

The potential at the peak of CV curves is termed the peak potential. For the reversible redox reaction, the anodic peak potential (E_{pa}) and cathodic peak potential (E_{pc}) are independent of scan rate and concentration, therefore the peak potentials supply information about the identity of the substrate species and the thermodynamic index for the oxidation/reduction of the studied species. In addition, the average value of the anodic and cathodic peak potentials corresponds to the formal reduction potential ($E^{o'}$) for a reversible couple (Figure 2.5).

On the other hand, when a relatively fast chemical reaction subsequent to electron transfer is involved in the electrode process, as indicated in Eqs. 2.4 and 2.5 (EC process), the peak potential shifts depending on the magnitude of its rate constant (k) and the size of the back peak in the reverse scan should be decreased. In this case the peak potentials cannot be used as the thermodynamic index for the oxidation/reduction of the studied species.

$$\mathrm{Red} \rightleftarrows \mathrm{Ox} + \mathrm{ne}^- \tag{2.4}$$

$$\mathrm{Ox} \xrightarrow{k} \mathrm{P} \tag{2.5}$$

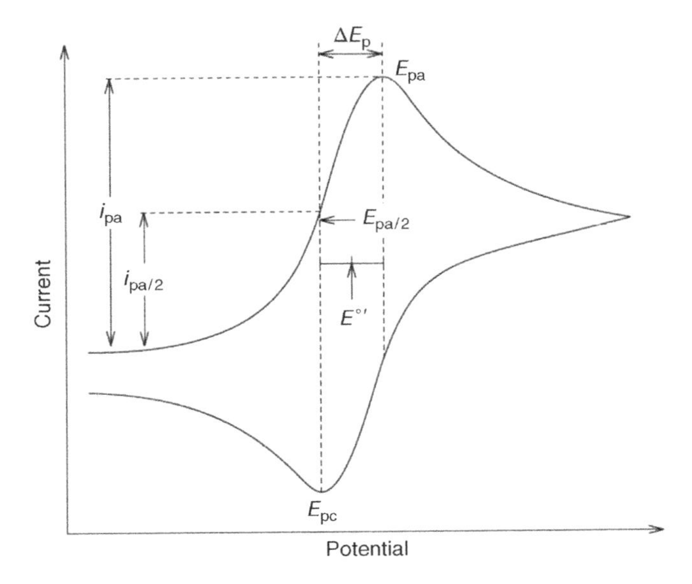

Figure 2.5 Cyclic voltammogram for a reversible process

The above discussion concerns voltammograms recorded at a fixed potential scan rate. However, if the scan rate is altered we are likely to observe a variation in the voltammograms recorded. For example, when cyclic voltammograms are recorded at a significantly faster scan rate (more than $1\,V\,s^{-1}$), a reversible response in CV may be observed due to the time taken to record the voltammogram. If the potential scan rate is sufficiently fast it is possible that no Ox formed according to Eq. 2.4 has had time to chemically react while the voltammogram is recorded. In this case, we can estimate a thermodynamic potential such as the formal potential from the obtained reversible voltammogram.

Although the half-peak potential ($E_{p/2}$) is often confused with the half-wave potential ($E_{1/2}$), it is defined as the potential where the current is half of the peak current in CV. In addition, the relationship between E_p and $E_{p/2}$ is represented by Eq. 2.6:

$$E_p - E_{p/2} = 0.057/n \tag{2.6}$$

where n is the number of electrons involved in the redox electrode reaction. However, $E_{p/2}$ is not an important thermodynamic parameter, and hence it is not used very much.

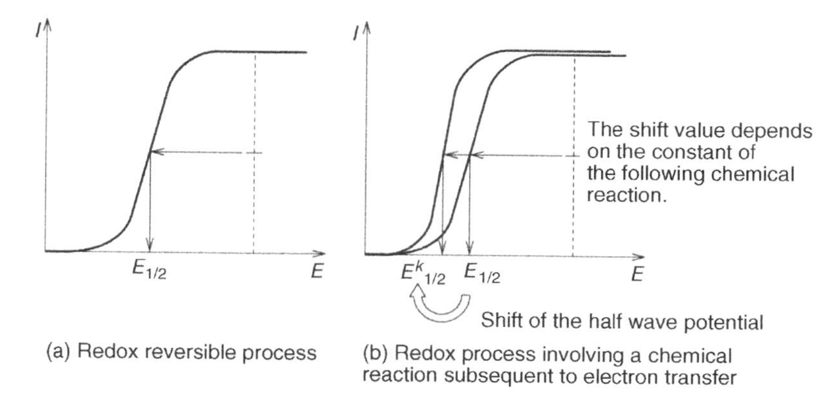

Figure 2.6 Steady-state voltammograms of a simple redox system (a) without and (b) with following chemical reaction

The half-wave potential ($E_{1/2}$) is the potential at which the wave current in steady-state voltammetry is equal to half of the diffusion limiting current (Figure 2.6a). $E_{1/2}$ is equal to the formal potential of the studied electrode reaction when the reaction is a redox reversible process and the diffusion coefficients of the oxidized and reduced forms of a studied species are equal to each other. However, if the reaction involves a chemical reaction subsequent to electron transfer, $E_{1/2}$ is no longer equal to the formal potential (Figure 2.6b).

The decomposition potential (E_{dec}) is a potential where the Faradic current begins to be observed on the voltammogram and is also called the onset potential (E_{onset}). Although E_{dec} is often used as the thermodynamic index for the oxidation/reduction of the studied species, it is not exactly a thermodynamic parameter.

2.8 UTILIZATION OF VOLTAMMETRY FOR THE STUDY OF ORGANIC ELECTROSYNTHESIS

2.8.1 Voltammetric Analysis for Selective Electrosynthesis

Many organic electrode processes are in principle multi-pathway, and therefore many reaction products are usually obtained under constant-current electrolysis. For example, constant-current electrolysis for the reduction of aromatic aldehydes or ketones under acidic conditions gives

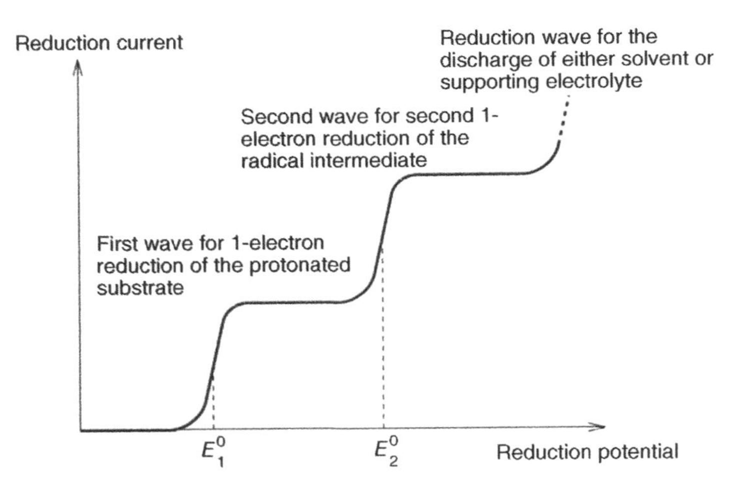

Figure 2.7 Linear sweep voltammogram for the reduction of an aromatic carbonyl compound under mechanical stirring

two products, such as alcohols and pinacols, simultaneously (Eq. 2.7) [10,11]. Consider linear sweep voltammetry (LSV) for this reduction under mechanical stirring (Figure 2.7). It is known that the first wave corresponds to one-electron reduction of the protonated substrate to form the corresponding radical intermediate, while the second wave corresponds to a second one-electron reduction of the radical intermediate to form the alcohol, as shown in Eq. 2.7. Therefore, when the potential of the working electrode is maintained at potential E^o_1 (known as the constant-potential electrolysis), one-electron reduction of the carbonyl compound will proceed clearly to the radical intermediate and consequently the pinacol product can be obtained selectively. If, on the other hand, the cathode is maintained at potential E^o_2, a two-electron reduction as well as the one-electron reduction will occur, and a mixture of two products is obtained.

$$\mathrm{Ar{-}C({=}O){-}R} \xrightarrow{\mathrm{H^+}} \mathrm{Ar{-}\overset{+}{C}(OH){-}R} \xrightarrow[E_1^{o}]{e^-} \mathrm{Ar{-}\overset{\bullet}{C}(OH){-}R} \xrightarrow{\text{Dimerization}} \tfrac{1}{2}\ \underset{\text{pinacol}}{\mathrm{ArCR(OH){-}CR(OH)Ar}}$$

$$\mathrm{Ar{-}\overset{\bullet}{C}(OH){-}R} \xrightarrow[E_2^{o}]{e^-} \mathrm{Ar{-}\overset{-}{C}(OH){-}R} \xrightarrow{\mathrm{H^+}} \underset{\text{alcohol}}{\mathrm{Ar{-}CHR(OH)}} \qquad (2.7)$$

Thus LSV in this process provides information about the reaction mechanism that allows us to use the process selectively in synthesis.

2.8.2 Clarification of the Reaction Mechanism

In general, organic electrode reactions include several varieties of coupled chemical reactions, and hence their mechanism is usually complex compared to that for inorganic electrode reactions. When we undertake a new organic electrode reaction it is therefore desirable to conduct CV measurements in advance in order to understand the mechanism.

A CV experiment conducted by Michielli and Elving is a good example for clarifying the complex reaction mechanism [12]. They employed wave clipping and addition of a proton donor (phenol) to obtain information about intermediates in the electrochemical reduction of benzophenone (**1**). As shown in Figure 2.8a, CV for the reduction gives two successive cathodic peaks I_c and II_c at −1.8 and −2.0 V (relative to the Ag/$AgNO_3$ reference electrode), and a single anodic peak I_a, as the potential is scanned from 0 to −2.3 V and back to −1.0 V. Wave clipping, that is, reversal of the scan direction at a potential (−1.9 V) between peaks I_c and II_c, causes peak Ia to remain, as shown in Figure 2.8b.This proves very nicely that peak I_a is associated with oxidation of the species formed in the first reduction step (I_c) and reduction at peak I_c therefore forms a species that is long lived on the CV time scale. On the other hand, peak II_c must

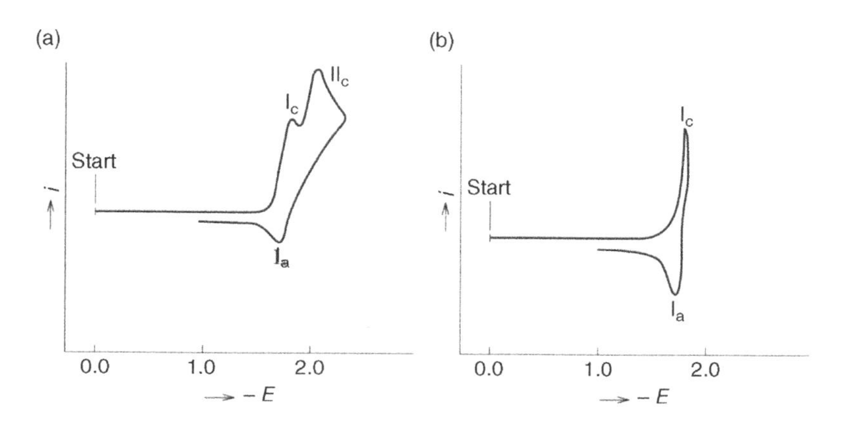

Figure 2.8 Cyclic voltammograms for the reduction of benzophenone. (a) Potential is scanned from 0 to −2.3 V and back to −1.0 V. (b) Potential is scanned from 0 to −1.9 V and back to −1.0 V

produce a very reactive species because it is totally irreversible. Addition of increasing amounts of phenol to the solution increased the height of peak I_c and diminished peaks II_c and I_a until finally peak I_c had doubled in size and peaks II_c and I_a were no longer evident.

These results are consistent with the following mechanism:

$$\underset{\mathbf{1}}{(C_6H_5)_2CO} + e^- \longrightarrow \underset{\mathbf{2}}{\left[(C_6H_5)_2CO\right]^{\dot{-}}} \tag{2.8}$$

$$\mathbf{2} + e^- \longrightarrow \underset{\mathbf{3}}{\left[(C_6H_5)_2CO\right]^{2-}} \tag{2.9}$$

$$C_6H_5OH + \mathbf{2} \longrightarrow C_6H_5O^- + \underset{\mathbf{4}}{(C_6H_5)_2\dot{C}OH} \tag{2.10}$$

$$\underset{\mathbf{3}}{\left[(C_6H_5)_2CO)\right]^{2-}} \xrightarrow{H^+} \underset{\mathbf{5}}{(C_6H_5)_2\bar{C}OH} \tag{2.11}$$

$$\mathbf{4} + e^- \longrightarrow \mathbf{5} \tag{2.12}$$

$$\mathbf{5} \xrightarrow{H^+} \underset{\mathbf{6}}{(C_6H_5)_2CHOH} \tag{2.13}$$

The couple I_c–I_a is associated with reversible formation of benzophenone ketyl **2** (Eq. 2.8). Peak II_c should be associated with reduction of **2** to the reactive dianion **3** (Eq. 2.9). In the presence of phenol, **2** is protonated (Eq. 2.10), and the resulting radical **4** is then further reduced to form the monoanion **5** (Eq. 2.12). Finally **5** is protonated to form the alcohol **6** (Eq. 2.13).

As we have seen in the above example, CV provides preliminary information about the mechanism of an unknown electrode reaction. To further clarify the mechanism, we can carry out the electrochemical reaction on a large enough scale to permit the products and intermediates to be isolated and identified. CV measurements of the products and intermediates isolated by preparative scale electrolysis may also help us to understand the mechanism of the studied electrochemical reaction. Thus, both CV measurement and preparative scale electrolysis are absolutely essential to clarify the complex mechanism of an unknown organic electrode reaction.

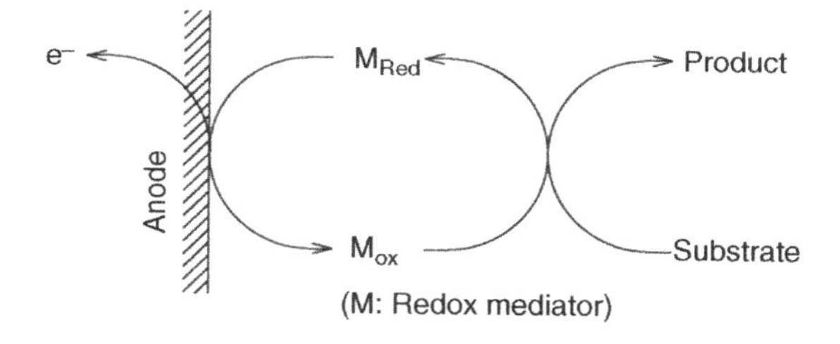

Figure 2.9 Principle of a redox mediatory reaction. This is for an oxidation process

2.8.3 Voltammetry for Selection of Mediator

Most organic electrode reactions are carried out by direct electrolysis, that is, by electron transfer between the organic substrate and the electrode. On the other hand, by adding a redox mediator to the medium it is possible to carry out electrochemical reactions even at a potential where the substance of interest is electroinactive. In this system, as shown in Figure 2.9 (this is for the oxidation process, but exactly analogous behaviour can be observed for the reduction process), electron interchange between the mediator M_{Red} (catalyst) and the electrode generates a substance M_{Ox} that can in turn undergo an electron interchange with the substrate molecule to give the product. This kind of system is called a mediated electrocatalytic reaction and the reaction becomes an indirect electrolysis [13].

The features of a mediated electrocatalytic reaction are as follows:

1. A catalytic amount of mediator is enough to carry out indirect electrolysis.
2. The substrate undergoes a redox conversion at a potential lower than that required to effective its direct electrolysis.
3. Passivation of the electrode as a result of the formation of non-conductive polymers during direct electrolysis can be avoided.
4. Highly selective and efficient redox conversion can be carried out.

Frequently used mediators for indirect electrolyses are multivalent metal ions, halide ions, polycyclic aromatic hydrocarbons such as naphthalene and anthracene, triarylamines and nitroxyl radicals. In addition, many mediators have been synthesized for specific indirect electrolyses.

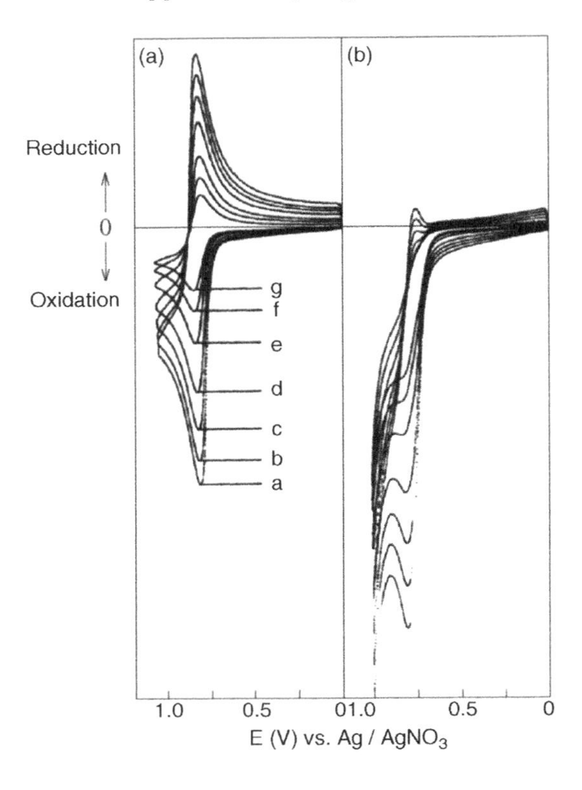

Figure 2.10 Cyclic voltammograms for the oxidation of tris-(*p*-bromophenyl)amine in the (a) absence and (b) presence of 4-methoxybenzyl alcohol

In this situation CV measurements are usually used to select a suitable mediator in the studied electrochemical reaction.

Since the direct electrochemical oxidation of alcohols to carbonyl compounds is carried out with great difficulty, oxidation has been a favourite testing ground for new electrocatalytic systems. Triarylamines are well known as excellent mediators for the indirect electrochemical oxidation of various alcohols. For example, tris-(*p*-bromophenyl)amine exhibits a typical reversible redox response in CV ($E^{o'} = 0.8$ V vs $Ag/AgNO_3$), as shown in Figure 2.10a [14]. On the other hand, by adding 4-methoxybenzyl alcohol as a substrate, the anodic peak increases while the cathodic peak becomes smaller or disappears (Figure 2.10b). This behaviour suggests that tris-(*p*-bromophenyl)amine can play the role of mediator and direct electrochemical oxidation of the substrate can take place smoothly. However, when benzyl alcohol is added to the tris-(*p*-bromophenyl)amine solution such mediatory behaviour is never observed in CV and a reversible redox response of the amine is maintained because the oxidation potential of benzyl alcohol ($E_{pa} = 2.0$ V vs $Ag/AgNO_3$) is much higher than that of 4-methoxybenzyl alcohol ($E_{pa} = 1.2$ V vs $Ag/AgNO_3$). To observe mediatory behaviour for benzyl alcohol oxidation, a more powerful mediator such as tris-(2,4-dibromophenyl)amine ($E^{o'} = 1.2$ V vs $Ag/AgNO_3$) is required. Thus CV study is a powerful tool for selecting a suitable mediator in this indirect electrochemical reaction.

2.8.4 Voltammetry for Selection of Electrode Material

As mentioned in section 2.1, the upper and lower limits of the potential window are determined by the working electrode material, and hence the selection of a working electrode material is critical not only to experimental success for voltammetry but also smooth progress of the desired preparative scale electrolysis.

Although the electrochemical oxidation of furan in methanol solution usually gives the corresponding methoxylated product, as shown in Eq. 2.14, the desired oxidation proceeds inefficiently at the Pt electrode. On the other hand, an excellent yield of product can be obtained by using a glassy carbon (GC) working electrode. The influence of the working electrode material on the efficiency of the electrochemical reaction can also be confirmed by the CV study [15].

– e
MeOH, – H+
O
OMe
– e
MeOH, – H+
OMe
MeO
OMe

(2.14)

Figures 2.11a and 2.11b show cyclic voltammograms for the oxidation of furan in methanol solution recorded at GC and Pt anodes, respectively. An oxidation peak for furan can be observed at a very positive potential of 1.9 V vs. SCE using the GC plate as the anode. In contrast, when the Pt plate is used as the anode material, methanol solvent is discharged more easily, a high background current is recorded and therefore no oxidation peak for the oxidation of furan is observed. It can therefore be stated from

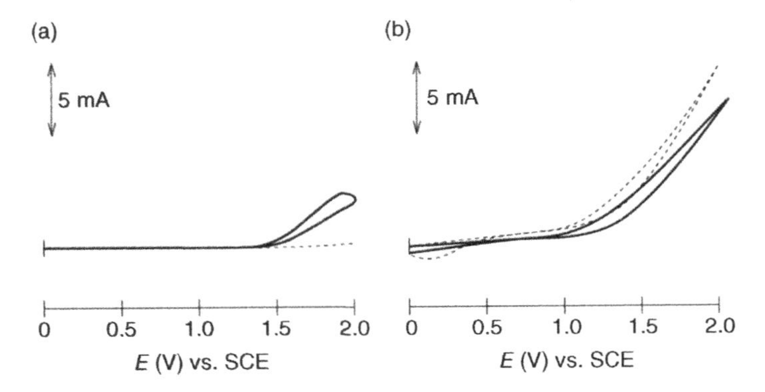

Figure 2.11 Cyclic voltammograms for the oxidation of furan in methanol solution recorded at (a) a GC plate anode and (b) a Pt plate anode. Dashed curves indicate the response in the absence of furan

the CV experiments that GC is the better anode material for the oxidation of furan.

As we have seen in the above example, voltammograms measured in this process provide preliminary information about a suitable electrode material that will allow us to use the process efficiently in synthesis.

REFERENCES

1. Bard, A.J. and Faulkner, L.R. (2001) *Electrochemical Methods, Fundamentals and Applications*, 2nd edn, John Wiley & Sons, Inc., New York, Chapter 2.
2. Izutsu, K. (2009) *Electrochemistry in Nonaqueous Solutions*, Wiley-VCH Verlag GmbH, Weinheim, Chapter 4.
3. Izutsu, K. (2009) *Electrochemistry in Nonaqueous Solutions*, Wiley-VCH Verlag GmbH, Weinheim, Chapter 6.
4. Bard, A.J. and Faulkner, L.R. (2001) *Electrochemical Methods, Fundamentals and Applications*, 2nd edn, John Wiley & Sons, Inc., New York, Chapter 1.
5. Izutsu, K. (2009) *Electrochemistry in Nonaqueous Solutions*, Wiley-VCH Verlag GmbH, Weinheim, Chapter 1.
6. Fry, A.J. (1989) *Synthetic Organic Electrochemistry*, John Wiley & Sons, New York, Chapter 10.
7. Lund, H. and Hammerich, O. (eds) (2001) *Organic Electrochemistry*, 4th edn, Marcel Dekker, New York, Chapter 1.
8. Fry, A.J. (1989) *Synthetic Organic Electrochemistry*, John Wiley & Sons, New York.
9. Compton, R.G. and Banks, C. (2000) *Understanding Voltammetry*, World Scientific Publishing, Singapore.

10. Cheng, P.-C. and Nonaka, T. (1989) *J. Electroanal. Chem.*, **269**, 223–230.
11. Atobe, M., Matsuda, K. and Nonaka, T. (1996) *Electroanalysis*, **8**, 784–788.
12. Michielli, R.F. and Elving, P.J. (1968) *J. Am. Chem. Soc.*, **90**, 1989–1995.
13. Fry, A.J. (1989) *Synthetic Organic Electrochemistry*, John Wiley & Sons, New York, Chapter 9.
14. Brinkhaus, K.-H.G., Steckhan, E. and Schmidt, W. (1983) *Acta Chem. Scand.*, **B37**, 499–507.
15. Horii, D., Atobe, M., Fuchigami, T. and Marken, F. (2006) *J. Electrochem. Soc.*, **153**, D143–D147.

3

Methods for Organic Electrosynthesis

Toshio Fuchigami

Organic electrosynthesis is generally affected by more complicated factors than ordinary organic synthesis therefore we must choose suitable electrolytic cells (divided or undivided), electrolytic methods (constant current or constant potential), electrodes, supporting electrolytes, solvents and so on [1–5]. In this chapter, the detail is described so that even beginners will be able to carry out organic electrosynthesis.

3.1 SELECTION OF ELECTROLYTIC CELLS

A proper choice of cell design is important in performing the desired electrolytic reaction. Organic electrolytic reactions are achieved on a laboratory scale by using an undivided cell. The simplest cell design is shown in Figure 3.1, but a cylindrical cell, shown in Figure 3.2, is the recommended design when anhydrous conditions or electrolysis under an inert gas atmosphere, like nitrogen, is required. In an inert gas atmosphere

Fundamentals and Applications of Organic Electrochemistry: Synthesis, Materials, Devices, First Edition. Toshio Fuchigami, Mahito Atobe and Shinsuke Inagi.

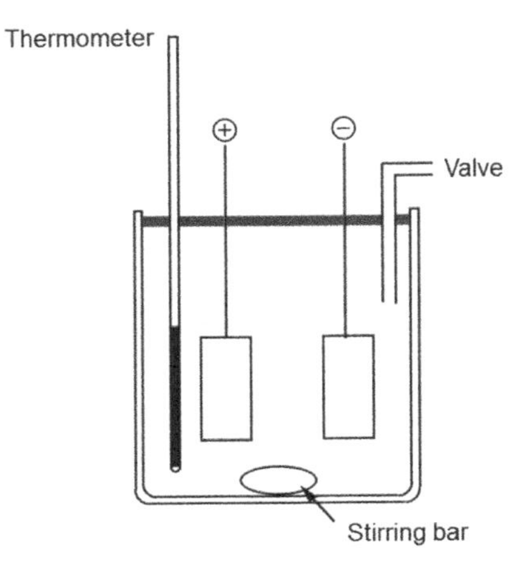

Figure 3.1 Beaker-type cell

the solvent and substrate are injected by syringe into the cell through a rubber septum.

When a species reduced at a cathode is oxidized at an anode, and vice versa, a two-compartment cell, namely a divided cell with a diaphragm (sintered glass or an ion-exchange membrane), should be used to prevent mixing of the anodic and cathodic solutions. An H-type cell divided with a sintered glass diaphragm (pore size $\varphi = 5\text{–}10\ \mu m$) is convenient, as shown

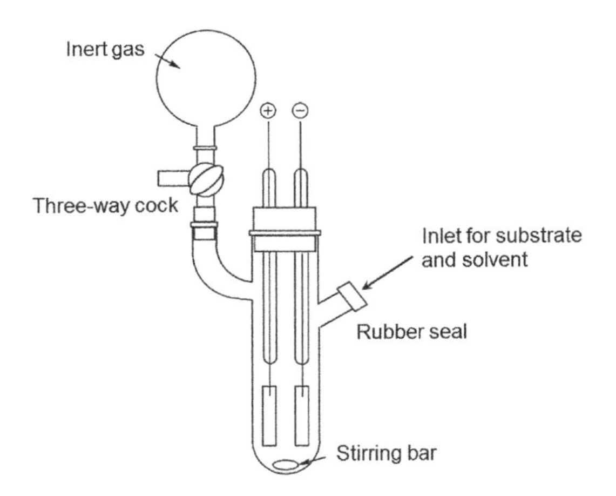

Figure 3.2 Undivided cell for anhydrous electrolysis

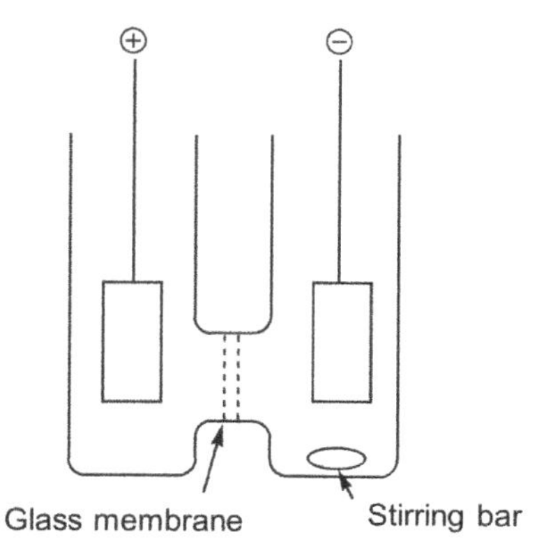

Figure 3.3 H-type cell with glass filter diaphragm

in Figure 3.3. A suitable volume for each compartment is 10–200 cm^3, and the diameter of the diaphragm should be as large as possible to decrease the cell resistance. When starting substrate and/or products migrate to a counter-compartment through a sintered glass diaphragm, an ion-exchange membrane should be used instead of the diaphragm. In this case, a separable cell in which the membrane is sandwiched between the compartments using screws is recommended. Such separable divided cells are commercially available. When a diaphragm is used, the cell resistance usually increases and consequently the cell voltage (the voltage between the anode and cathode) increases. To decrease the cell voltage, the distance between the electrodes should be kept as small as possible.

Even when a divided cell is necessary for cathodic reduction, we can achieve the desired reduction using an undivided cell as follows. Use of sacrificial anodes like Mg and Zn or addition of sacrificial organic compounds like oxalic acid and acetic acid to the electrolytic solution enables re-oxidation of cathodic products to be avoided. In the former case the anode metals are dissolved as metal ions in the solution during the electrolysis, while in the latter case the acids are anodically oxidized to form CO_2 and ethane. Such sacrificial electrolytic systems enable oxidation of cathodic products and the starting material at the anode to be avoided. On the other hand, in the case of anodic oxidation requiring a divided cell, a proton source should be added as a sacrificial substance into the electrolytic solution and cathode materials with low hydrogen overpotential should be used. In this case, proton reduction at the cathode takes place preferentially, which enables the reduction of anodic

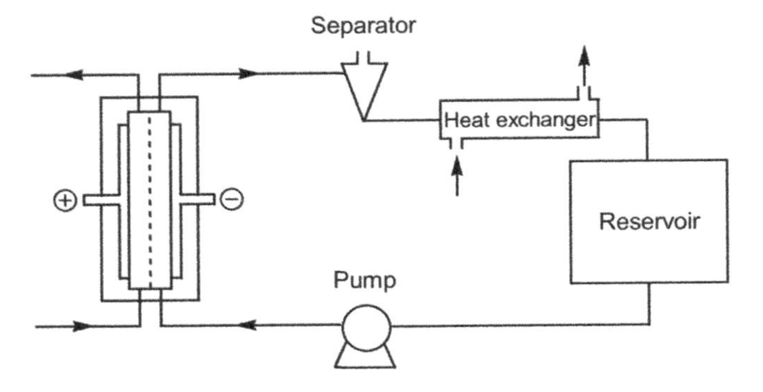

Figure 3.4 Filter press type flow cell

products and the starting material at the cathode to be avoided. Consequently, we can achieve the desired anodic oxidation selectively even in an undivided cell.

Crucibles made of conducting materials like graphite and glassy carbon can be used as working electrodes. In this case, the volume of electrolytic solution can be reduced and the surface area of the working electrode can be increased.

For scaling up for commercialization a filter press type flow cell is recommended, as shown in Figure 3.4. In this case, the volume of the electrolytic solution has no limit since the solution is circulated by pump, and mass transport from the bulk to the surface of the working electrode is promoted to increase the efficiency of the electrolytic reaction.

3.2 CONSTANT CURRENT ELECTROLYSIS AND CONSTANT POTENTIAL ELECTROLYSIS

It is recommended that organic electrosynthesis be carried out at a constant current at first because the setup of the electrolytic system and the operation of the power supply are simple. The product selectivity and yield can be improved by controlling current density and the amount of electricity passed. Since current density is correlated to applied potential, changing the current density creates a potential shift. The current density should be selected depending on the concentration of the starting substrate. When the concentration of the substrate is low, the current density should be low and vice versa. The electricity passed is readily

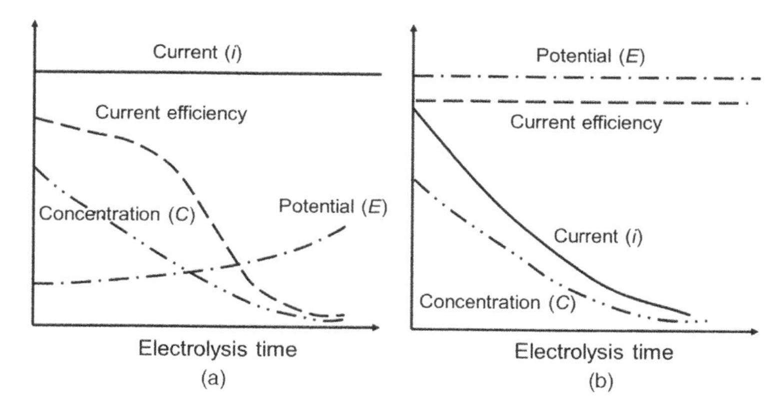

Figure 3.5 Profile showing electrolysis time during constant current electrolysis (a) and constant potential electrolysis (b)

calculated according to the following equation: current $(A) \times$ time $(s) =$ electricity (C). For instance, when the desired electrolytic reaction is a two-electron reaction, the theoretical amount of electricity is 2 F (2×96480 C). When this electricity is divided by applied current (A), one can easily calculate for how many hours the electrolysis must be carried out. As shown in Figure 3.5a, the electrode potential changes with the consumption of the starting substrate (positive shift in case of oxidation or negative shift in case of reduction), therefore the product selectivity and current efficiency sometimes decrease, particularly in the late stage of electrolysis. However, highly selective and efficient organic electrosynthesis can often be achieved even at constant current electrolysis, hence commercialized electrode processes are operated mainly by constant current electrolysis. Nevertheless, a constant potential electrolysis is suitable for achieving high selectivity and clarification of the reaction mechanisms. Moreover, based on constant potential coulometry, one can easily find the number of electrons involved in the electrolytic reaction. In order to carry out constant potential electrolysis, the oxidation potential or reduction potential of the substrate should be measured in advance, but one can estimate the applied potential at which rapid current increase is observed from a current-potential curve of the substrate. As shown in Figure 3.6, a salt bridge terminated either by a Luggin capillary or a plug of porous Vycor glass is placed closed to the working electrode, and an appropriate constant potential relative to a reference electrode such as an aqueous SCE is applied using a potentiostat. The amount of electricity passed is measured by a coulometer.

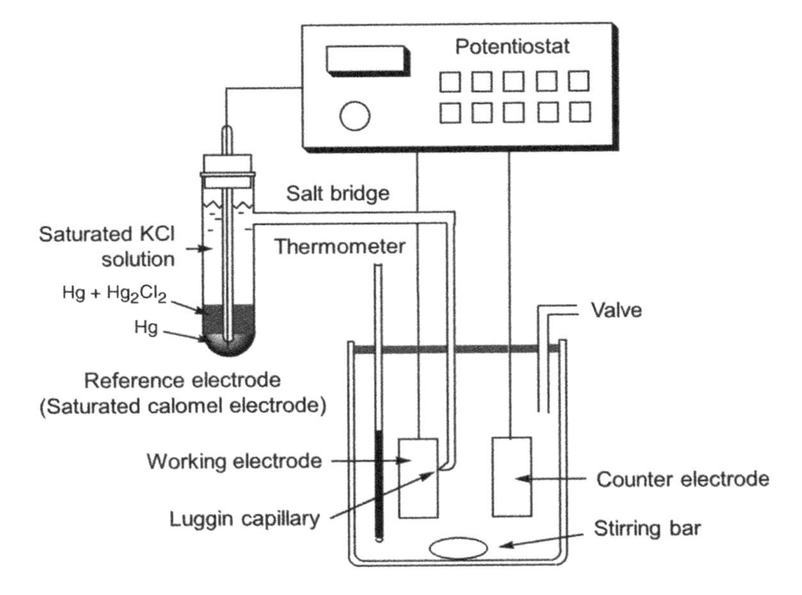

Figure 3.6 Setup for constant potential electrolysis

3.3 DIRECT ELECTROLYSIS AND INDIRECT ELECTROLYSIS

Electrolysis is classified as direct electrolysis or indirect electrolysis. The former is based on direct electron transfer between substrate molecule and electrode, which is a simple and common electrolytic method. The latter is based on electron transfer using redox mediators dissolved in an electrolytic solvent, as shown in Figure 3.7 (indirect cathodic reduction). (Note that indirect anodic oxidation is covered in Chapter 2, Figure 2.9.) Indirect electrolysis is also classified into in-cell, where electrolysis is carried out in

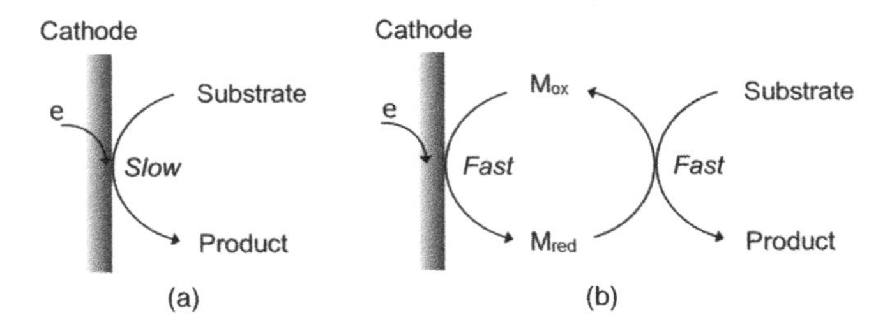

Figure 3.7 Principle of indirect electrolysis using a mediator

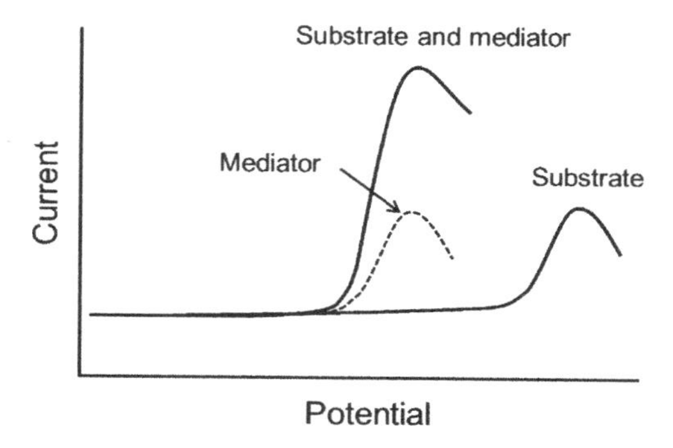

Figure 3.8 Current potential curve for direct and indirect electrolysis

the presence of both substrate and mediator, and ex-cell, where electrolysis is used for only regeneration of the mediator. In the former case, the redox potential of the mediator should be lower than that of the substrate. When the redox potential of the mediator is higher than that of the substrate, the latter type is employed. In the former case a catalytic amount of mediator should be enough, while in the latter case a quantitative or excess amount of mediator is necessary. When the heterogeneous electron transfer between mediator and electrode, as well as the redox reaction with the substrate, is fast enough (Figure 3.7b), a significantly enhanced catalytic current can be obtained due to decreased electrolysis potential (decreased activation energy), as shown in Figure 3.8. In other word, a large electrocatalytic current can be obtained.

The characteristics and proper choice of mediators and their synthetic applications are described in Chapters 2 and 4, respectively. Many types of mediators are available and the redox potentials of mediators can be tuned by their molecular design. Moreover, some mediators can be applied to electrochemical asymmetric synthesis [6–9].

3.4 ELECTRODE MATERIALS AND REFERENCE ELECTRODES

The choice of electrode material is one of the most important factors in electrolysis since the material is not only the interface for electron transfer with the substrate molecule but also acts as an electrocatalyst. If the correct electrode material is not chosen, the desired electrolytic reaction does not proceed, therefore a suitable electrode material must be selected carefully. In an aqueous solution, oxygen gas and hydrogen gas are generated competitively due to discharge of water during the electrolytic reaction. In order to avoid such undesirable reactions, cathode materials with high hydrogen overpotential or anode materials with high oxygen overpotential should be used. The order of oxygen overpotential is Au > Pt, Pd, Cd, Ag > PbO_2 > Cu > Fe > Co > Ni, and the order of hydrogen overvoltage is Hg > Zn, Pb, Cd > graphite > Cu > Fe, Ni > Ag, Co > Pt, Pd. A mercury cathode was very often used for electroreductive organic synthesis, but it is no longer used except for electrochemical measurements because of its toxicity. A platinum electrode is the first choice in aprotic organic solvents. Although various kinds of cathode materials are available, usable anode materials are limited. Most metals are easily oxidized apart from noble metals like platinum and gold. Carbon, graphite and metal oxides like PbO_2 are commonly used as anodes. Carbon electrodes generally include a trace amount of metals like iron, therefore their surface is paramagnetic. Hence, such carbon anodes easily capture anodically generated radicals to enhance further oxidation to form cationic intermediates. On the other hand, platinum anodes generally tend to generate radical intermediates selectively. Accordingly, platinum anodes are most suitable for Kolbe electrolysis. It should be noted that when a carbon anode is used in a solution containing $LiClO_4$, collapse of the electrode surface often takes place due to the oxidation of binder such as coal tar pitch in the carbon electrode. In sharp contrast, glassy carbon (GC) electrodes have enough durability for anodic oxidation, and anode passivation does not occur, which is quite different from other carbon electrodes. However, GC electrodes are costly, as are platinum electrodes. Although non-precious-metal electrodes are not suitable for anode materials, Mg, Zn, Al and Cu are often used as reactive electrodes or sacrificial anodes, as explained in section 3.1.

The following points should be noted when organic electrosynthesis is carried out. In contrast to electrochemical measurements, electrochemical synthesis requires a relatively large current of 10–100 mA ($\sim$10 mA cm^{-2}) even on a laboratory scale. The use of a larger electrode surface area is therefore recommended to complete the electrolysis in a shorter time. As working electrode shape, plate, candy stick (baculiform) and mesh type are commonly used. The surface area of the counter electrode should be same as that of the working electrode. Depending on the kind of electrodes, pre-treatment of the electrode surface differs, but in general the electrode surface is polished with alumina abrasive or emery papers.

Reference electrodes are required when measurement of current potential curves and constant potential electrolysis are carried out. When aqueous and protic solvents are used as electrolytic solvents, a SCE (Figure 2.2a) and a silver–silver chloride (Ag/AgCl) electrode (Figure 2.2b) should be used. For electrolysis in non-aqueous solution, an Ag/Ag^+ reference electrode is convenient since it can be placed near the working electrode surface without a salt bridge. Reference electrodes are described in further detail in Chapter 2.

3.5 ELECTROLYTIC SOLVENTS AND SUPPORTING ELECTROLYTES

Electrolytic solvents must have polarity, which allows supporting electrolytes to be dissolved to dissociate ions and provide sufficient ion conductivity for the electrolytic solutions. The index of solvent polarity is reflected by its dielectric constant, and electrolytic solvents should have low viscosity and toxicity. In particular, it should be noted that solvent viscosity strongly affects diffusion of substrate from bulk to electrode surface.

3.6 STIRRING

Since electrolytic reactions are typically heterogeneous, stirring of an electrolytic solution controls mass transport of substrate, which greatly affects current efficiency and product selectivity. Even if a solution is vigorously stirred, the transport of the substrate from bulk to the electrode surface is not always enhanced. As shown in Figure 3.4, circulation of an electrolytic solution through a flow cell, particularly a filter press type cell, is much more efficient compared to a conventional beaker type cell, hence the filter press type cell is usually employed in industrial electrolysis.

3.7 TRACKING OF REACTANT AND PRODUCT

Efficient conversion of substrate to desired product is very important in organic synthesis. In the case of constant current electrolysis, discharge of solvent and/or supporting electrolyte takes place simultaneously at a later stage of the electrolysis. A theoretical amount of charge passed is not always enough for complete substrate conversion, hence monitoring electrolysis (consumption of starting substrate and amount of product) should be carried out using TLC, GC, HPLC or mass spectrometry (MS) to confirm substrate conversion.

3.8 WORK-UP, ISOLATION AND DETERMINATION OF PRODUCTS

Work-up of electrolysis is similar to ordinary organic synthesis, but the following points should be noted.

1. Since the anolyte becomes acidic and the catholyte becomes basic after electrolysis using a divided cell, neutralization of the desired electrolytic solution should be done.
2. Since a large amount of supporting electrolyte is contained in the electrolytic solution, its removal should be done first. For example, when the electrolytic solvent is water soluble, the solvent must be removed by evaporation under reduced pressure. The remaining residue or water-insoluble electrolytic solution is mixed with water or brine (saturated NaCl aqueous solution), if necessary, neutralization is performed and the product is extracted with appropriate organic solvents like diethyl ether. It should be noted that some hydrophobic supporting electrolytes like Bu_4NBF_4 are insoluble in water, but soluble in organic solvents like CH_2Cl_2.

Isolation and identification techniques are similar to those in ordinary organic synthesis. Product isolation is carried out using distillation, recrystallization or various types of chromatography, and identification is performed using various spectroscopic analysis methods such as ^1H NMR, IR and MS.

3.9 CURRENT EFFICIENCY AND EFFECT OF THE POWER UNIT

Current efficiency is the most important factor for evaluating the results of electrolytic reactions. For instance, consider the following two-electron reduction of acetone:

$$(H_3C)_2C{=}O \xrightarrow{2e,\ 2H^+} (H_3C)_2CH{-}OH \qquad (3.1)$$

Electricity of 2 × 96,480 C is required for the formation of 1 mol of isopropyl alcohol. When n mol of isopropyl alcohol is formed from 1 mol of acetone after passing Q coulombs, the current efficiency (%) is $(96{,}480 \times 2n/Q) \times 100$. Current efficiency is usually below 100% since the solvent and/or supporting salt are discharged simultaneously during the electrolytic reaction.

The energy consumption for electrosynthesis is usually disregarded in basic research, but it is the most important factor in industrial electrolytic processes. The energy consumption for the production of the desired organic compound is shown as kWh kg^{-1}, which greatly depends on cell voltage, therefore it is recommended that the distance between electrodes should be kept as small as possible to decrease cell resistance.

REFERENCES

1. Bard, A.J. and Stratmann, M. (eds) (2007) Electrochemistry applied to organic synthesis: Principles and main achievements, in *Encyclopedia of Electrochemistry*, Vol. 5 (eds D. D. Macdonald and P. Schmuki), Wiley-VCH Verlag GmbH, Weinheim, Chapter 6.
2. Bard, A.J. and Stratmann, M. (eds) (2002) Methods to investigate mechanisms of electroorganic reactions (Chapter 1) and Practical aspects of preparative scale electrolysis (Chapter 2), in *Encyclopedia of Electrochemistry*, Vol. 8 (ed. H. J. Schäfer), Wiley-VCH Verlag GmbH, Weinheim.
3. (a) Lund, H. and Hammerich, O. (eds) (2001) *Organic Electrochemistry*, 4th edn, Marcel Dekker, Inc., New York, Chapters 1 and 2. (b) Hammerich, O. and Speiser, B. (eds) (2014) *Organic Electrochemistry*, 5th edn, CRC/Taylor & Francis.
4. Grimshaw, J. (2000) *Electrochemical Reactions and Mechanisms in Organic Chemistry*, Elsevier, Amsterdam.
5. Fry, A.J. (1989) *Synthetic Organic Electrochemistry*, Wiley Interscience, New York.
6. Tanaka, H., Kawakami, Y., Goto, K. and Kuroboshi, M. (2001) *Tetrahedron Lett.*, **42**, 445–448.
7. Kashiwagi, Y., Kurashima, F., Chiba, S., Anzai, J., Osa, T. and Bobitt, T.M. (2003) *Chem. Commun.*, 1124–1125.
8. Shiigi, H., Tanaka, H., Demizu, Y. and Onomura, O. (2008) *Tetrahedron Lett.*, **49**, 5247–5251.
9. Demizu, Y., Shiigi, H., Mori, H., Matsumoto, K. and Onomura, O. (2008) *Tetrahedron Asymmetry*, **19**, 2659–2665.

4

Organic Electrode Reactions

Toshio Fuchigami

In this chapter, the general features of electrolytic reactions and differences between electrolytic reactions and ordinary chemical ones will be explained. Typical characteristics such as umpolung (polarity inversion) and selectivity control together with electron transfer control are discussed. Reaction types and various electrochemically generated reactive species will also be explained [1–6].

4.1 GENERAL CHARACTERISTICS OF ELECTRODE REACTIONS

Organic electrode reactions have the following features, many of which cannot be achieved by other methods:

1. Electrode reactions are typically heterogeneous reactions, and the reaction fields are specific since oxidation and reduction take place separately at different fields.
2. Umpolung (polarity inversion) is readily performed without the use of any reagents.

Fundamentals and Applications of Organic Electrochemistry: Synthesis, Materials, Devices, First Edition. Toshio Fuchigami, Mahito Atobe and Shinsuke Inagi.

3. The selectivity of electrode reactions is often different from that of ordinary organic reactions.
4. Since electrons are used as a reagent, the use of hazardous reagents can be avoided, i.e. electrode reactions are low-emission processes.
5. Electrode reactions proceed under mild conditions such as room temperature and normal pressure.
6. Electrode reactions can be started or stopped readily by on-off switch of the power supply, i.e. electrode reaction control is easy.
7. The scale effect is generally small.

Electrode reactions are redox reactions through electron transfer between a substrate molecule and an electrode. The main reaction field is an electrode surface (solid–liquid interface) or near the electrode surface, and the surface has an extremely large electrical field, which is quite different from ordinary redox reactions on heterogeneous catalysts. Electrode reactions therefore take place in highly unique fields.

Using a chemical redox reaction as an example, the difference between an ordinary chemical reaction and an electrode reaction will be explained in detail. Figure 4.1a shows the reduction of substrate **B** by reducing reagent **A**. When an activated complex is formed or **A** and **B** approach each other closely enough for electron transfer, electron transfer from **A** to **B** takes place. Next, reductant **A** is transformed to oxidized **C** while substrate **B** is transformed to reduced product **D**. The former electron

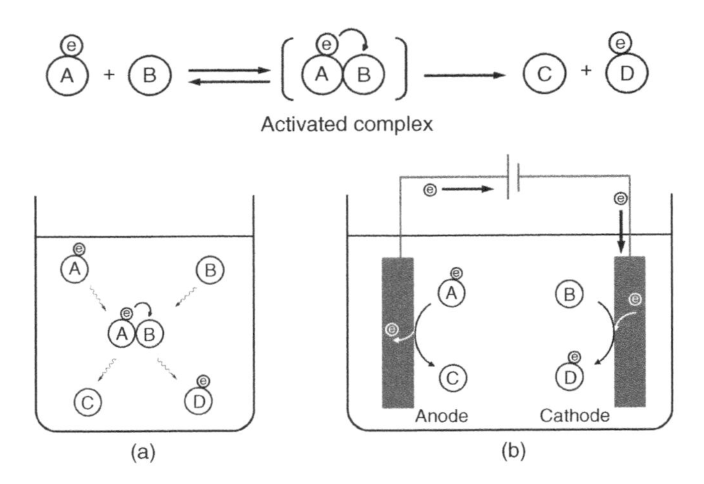

Figure 4.1 Difference between chemical reaction (a) and electrochemical reaction (b)

transfer through the complex is called inner-sphere electron transfer or bonded electron transfer, while the latter one without any complex is called outer-sphere electron transfer or non-bonded electron transfer. Thus, both oxidation and reduction occur at the same place in the case of an ordinary chemical reaction, while oxidation and reduction occur at different places, such as an anode and cathode, respectively, in the case of an electrochemical reaction (Figure 4.1b). In other words, electrode electron transfer takes place separately due to the existence of the electrode interfaces, which is a significant characteristic feature that is different from ordinary chemical reactions.

4.2 MECHANISM OF ORGANIC ELECTRODE REACTIONS

In the case of organic electrode reactions, electron transfer generally does not take place cleanly, and pre and/or post reactions usually accompany the transfer. This is quite different from inorganic electrode reactions. An organic electrode reaction consists of an electron transfer step as well as several chemical and physical steps.

Figure 4.2 illustrates each elementary reaction step of substrate S forming product P via intermediate I. In step (a), mass transport of S

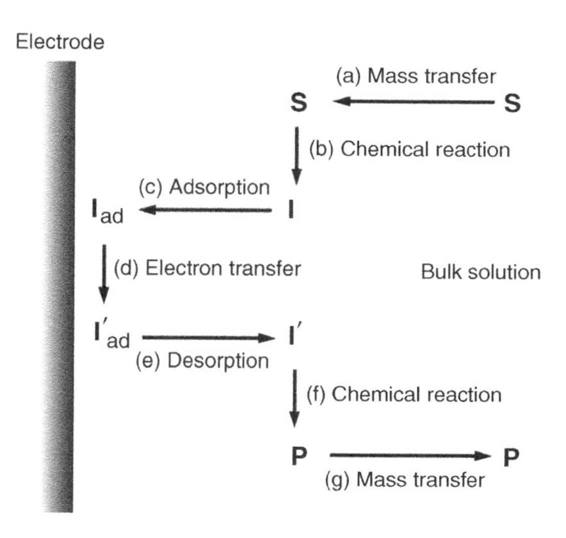

Figure 4.2 Elementary processes of electrode reactions

from the bulk of an electrolytic solution to the electrode surface takes place by diffusion or migration. In step (b), pre-reactions such as desolvation, dissociation and/or deprotonation of **S** take place to form intermediate **I**. However, such pre-reactions do not always take place. In step (c), the intermediate **I** adsorbs on the surface of the electrode to form intermediate $\mathbf{I}_{ad}$. In step (d), electron transfer between $\mathbf{I}_{ad}$ and an electrode generates intermediate $\mathbf{I}'_{ad}$. In step (e), desorption of $\mathbf{I}'_{ad}$ followed by subsequent chemical reaction (f) proceeds to provide a product **P** that diffuses to the bulk of the electrolytic solution, and then the sequential reaction is completed. Intermediate **I** may undergo an electron transfer reaction without an adsorption step (c) and also the order of sequential steps (e) and (f) may be reversed, i.e. $\mathbf{I}'_{ad}$ undergoes subsequent reaction and then desorption of the resulting product occurs. Thus, the electrode reaction is typical in a heterogeneous system, and mass transfer steps (a) and (g) as well as adsorption and desorption steps (c) and (e) are involved, which is quite different from homogeneous reactions.

If the electrode process (electron transfer process) is abbreviated to E and the chemical process is abbreviated to C, the organic electrode reaction can be shown using these abbreviations. For example, the electrochemical reaction illustrated in Figure 4.2 can be shown by the sequence CEC (pre-chemical reaction → electrode process → follow-up chemical reaction), and adsorption and desorption steps are usually disregarded unless they are important.

In order to clarify the reaction mechanism, electrochemical analyses such as coulometry and voltammetry in addition to ordinary organic mechanistic studies are necessary to obtain information such as the number of electrons transferred, redox potentials and detection of the reaction intermediates. The details of the electrochemical analyses are described in Chapters 2 and 3.

4.3 CHARACTERISTICS OF ORGANIC ELECTROLYTIC REACTIONS

4.3.1 Umpolung

The polarity inversion of chemical bonding can be readily carried out in electrolytic reactions [6]. In other words, electrophiles can be converted electrochemically to nucleophiles without use of any reagents, and vice versa. Such polarity inversion is widely used for organic synthesis. For example, alkyl halides are inherently electrophilic reagents. In order to convert them to nucleophilic reagents, they have to be transformed to Grignard reagents or lithium compounds using Mg or Li metal. However, such polarity inversion (umpolung) can be readily achieved in one step by cathodic reduction of the alkyl halides, as shown in Eq. 4.1. When the basic products are isolated after Grignard reaction, the work up can be troublesome, and it may be rather difficult to separate the product from aqueous alkaline solution due to insoluble $Mg(OH)_2$. In sharp contrast, in the case of electrolytic reactions, the product isolation is rather easy and the severe waste problem does not occur because there is no use of metals. In addition, electrolytic reactions do not require easily flammable ethereal solvents, and alternative solvents like acetonitrile can be used as electrolytic solvents.

$$R\text{-}X \begin{cases} \xrightarrow{2e,\,-X^-} R^- \xrightarrow{E^+} \\ \xrightarrow{Mg} RMgX \xrightarrow{E^+} \end{cases} R\text{-}E \tag{4.1}$$

When alkyl substituents are introduced to phosphine, silicon and stannum compounds, their corresponding chloro compounds (R_3ZCl, R_2PCl) and alkyl Grignard reagents are usually used. However, electrochemical reduction of these chloro compounds generates the anionic intermediates, which can readily react with alkyl halides (R′X) to provide the corresponding alkyl-substituted products (Eq. 4.2).

$$R_3Z-R' \xleftarrow{R'MgX} R_3Z-X \xrightarrow{2e,-X^-} R_3Z^- \xrightarrow{R'X} R_3Z-R' \quad (Z = Si, Sn)$$

$$R_2P-R' \xleftarrow{R'MgX} R_2P-X \xrightarrow{2e,-X^-} R_2P^- \xrightarrow{R'X} R_2P-R' \tag{4.2}$$

Allylsilanes usually react with electrophiles, but their electrochemical oxidation generates allylic cations, which react with nucleophiles as shown in Eq. 4.3.

$$CH_2{=}CHCH_2E \xleftarrow[-Me_3Si^+]{E^+} CH_2{=}CHCH_2SiMe_3 \xrightarrow[-Me_3Si^+]{-2e} [CH_2{\cdots}CH{\cdots}CH_2]^+ \xrightarrow{Nu^-} CH_2{=}CHCH_2Nu \quad (Nu = OR, OAc) \tag{4.3}$$

Electron-rich benzene derivatives usually react with electrophiles, while they can react with nucleophiles by electrochemical oxidation (Eq. 4.4).

$$\text{Ph-R} \xrightarrow{-e} [\text{Ph-R}]^{\bullet+} \xrightarrow{Nu^-} \text{(H)(Nu)C}_6\text{H}_5\text{-R}^{\bullet} \xrightarrow{-e} \text{(H)(Nu)C}_6\text{H}_5\text{-R}^{+} \xrightarrow{-H^+} \text{Nu-C}_6\text{H}_4\text{-R} \quad (4.4)$$

Furthermore, industrialized electroreductive hydrodimerization of acrylonitrile, shown in Eq. 4.5, is also a typical example of umpolung using cathodic reduction.

$$CH_2{=}CHCN \xrightarrow{e} [CH_2{=}CHCN]^{\bullet-} \xrightarrow{CH_2=CHCN} \xrightarrow{e,\ 2H^+} NC(CH_2)_4CN \quad (4.5)$$

4.3.2 Selectivity

The selectivity of electrolytic reactions is rather complicated since it is controlled by many factors, such as electrode materials, applied potential, current density, electrolytic solvents, supporting electrolytes, electric field, adsorption orientation of substrate or intermediate species at the electrode surface and so on [4–7]. Heterogeneous electrolytic reactions therefore often exhibit different selectivity from ordinary homogeneous chemical reactions. In particular, when an electrochemically generated reactive species or intermediate reacts with reagents before it diffuses from the electrode into the solution, the stereo- and regioselectivities of the product are often quite different from those of ordinary chemical reactions. Typical examples of the selectivity of electrode reactions are described in the following sections.

4.3.2.1 Chemoselectivity

Chemoselectivity in ordinary chemical reactions is controlled by the choice of reagents, but it is quite difficult to achieve high chemoselectivity when multiple similar functional groups exist in a single molecule. On the other hand, chemoselectivity can be achieved by the control of applied potential based on the difference between the redox potentials of functional groups. Since the applied potential is adjusted precisely using a potentiostat, high selectivity can be readily achieved. For instance, even though phenylimino and alkylimino groups exist in the same molecule, as shown in Eq. 4.6, the phenylimino group can be predominantly reduced by constant potential electrolysis. This is because the phenylimino group is more easily reduced than the alkylimino group as a result of the electron-withdrawing phenyl group. However, such selective reduction cannot be achieved by ordinary reducing reagents.

NR NPh —(2e, $2H^+$)→ NR NHPh (4.6)

Similarly, in the case of the molecule with three halogen atoms shown in Eq. 4.7, the halogen atom at the α-position to the carbonyl group is most easily reducible, thereby this halogen can be predominantly reduced at constant potential electrolysis.

—(2e, H^+)→ 94% (4.7)

4.3.2.2 *Reaction Pathway Selectivity*

It is known that the reaction pathway is greatly changed depending on applied potentials. As shown in Eq. 4.8, one-electron and two-electron reduction products are obtained selectively depending on the applied cathode potentials.

e, −1.2 V; 2e, −1.8 V; H^+ (4.8)

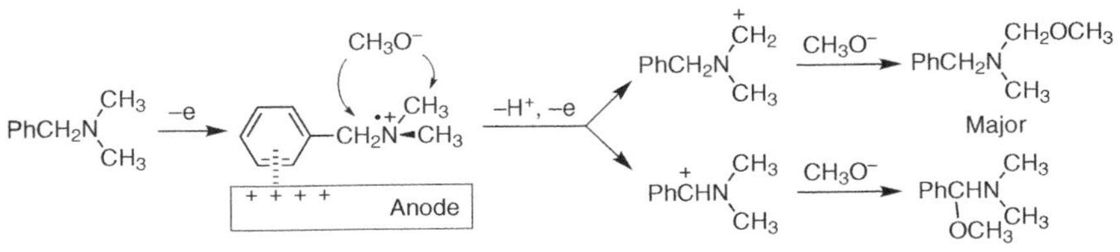

Figure 4.3 Regioselective anodic methoxylation

4.3.2.3 Regioselectivity

(a) Kinetically Controlled and Thermodynamically Controlled Regioselectivity Regioselectivity is often controlled by either the stability of a reactive intermediate (thermodynamic control) or the reaction rate of an intermediate with a reagent as well as the elimination rate of a leaving group (kinetic control). As a typical example of the former, it is well known that anodic benzylic substitutions easily take place and are attributable to a stable benzylic cation intermediate.

On the other hand, when *N*,*N*-dimethylbezylamine is anodically oxidized in methanol, a methoxy group is selectively introduced to the methyl group, as shown in Figure 4.3 [8]. This regioselectivity is not controlled by the stability of the cationic intermediate, i.e. this reaction is not thermodynamically controlled. Since the anodically generated radical cation intermediate seems to be adsorbed on the anode, deprotonation takes place preferentially from a less hindered methyl group. Consequently, this reaction is kinetically controlled.

In a similar manner, anodic methoxylation of *N*-ethyl-*N*-methylaniline, the corresponding carbamate, and amide derivatives also takes place at the methyl group selectively (Eq. 4.9) [9].

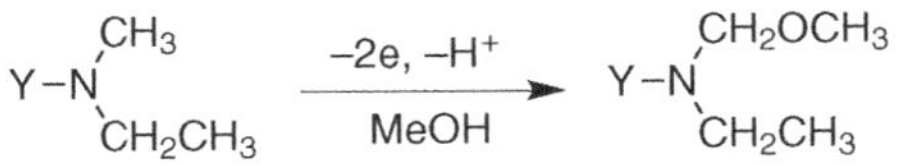

$$\text{Y-N(CH}_3\text{)(CH}_2\text{CH}_3\text{)} \xrightarrow[\text{MeOH}]{-2e,\ -H^+} \text{Y-N(CH}_2\text{OCH}_3\text{)(CH}_2\text{CH}_3\text{)} \tag{4.9}$$

(Y = Ph, ROCO, RCO)

However, the regioselectivity can also be explained by the difference in deprotonation rates of radical cation intermediates (so-called kinetic acidity [10]). For example, in the case of anodic methoxylation of an aniline derivative having a *N*-fluoroalkyl (Rf) group and an *N*-alkyl group, as shown in Figure 4.4, methoxylation takes place at the adjacent

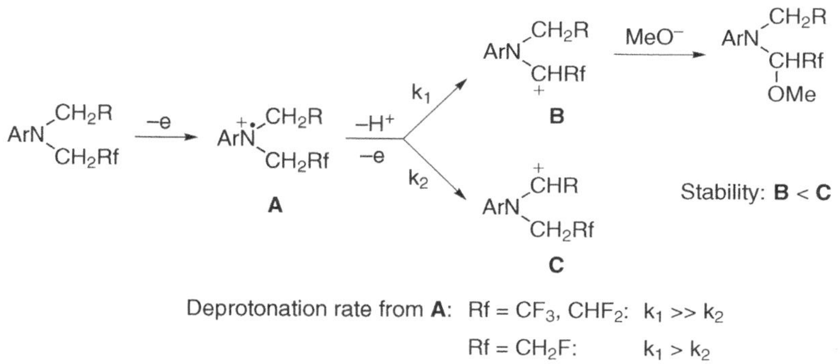

ArN(CH_2R)(CH_2Rf) $\xrightarrow{-e}$ $ArN^{+\bullet}$(CH_2R)(CH_2Rf) **A** $\xrightarrow{-H^+,\ -e}$

k_1: ArN(CH_2R)($\overset{+}{C}HRf$) **B** $\xrightarrow{MeO^-}$ ArN(CH_2R)(CHRf–OMe)

k_2: ArN($\overset{+}{C}HR$)(CH_2Rf) **C**

Stability: **B** < **C**

Deprotonation rate from **A**: Rf = CF_3, CHF_2: $k_1 >> k_2$

Rf = CH_2F: $k_1 > k_2$

Figure 4.4 Mechanism of regioselective anodic methoxylation (kinetic control)

position to the Rf group preferentially. The regioselectivity increases with an increase in the electron-withdrawing ability of the Rf group, i.e. the selectivity increases in the following order: $CH_3 < CH_2F < CHF_2 < CF_3$ [9]. The mechanism for this regioselectivity outcome is called kinetic control.

The regioselectivity of electrolytic fluorination is also controlled by kinetic acidity. For example, in the case of electrochemical fluorination of heterocyclic compounds, as shown in Eq. 4.10, fluorination proceeds predominantly via the unstable cation intermediate adjacent to the carbonyl group rather than via the stable benzylic cation. This can be explained in terms of enhanced facile deprotonation of the anodically generated radical cation by an electron-withdrawing carbonyl group [10–12].

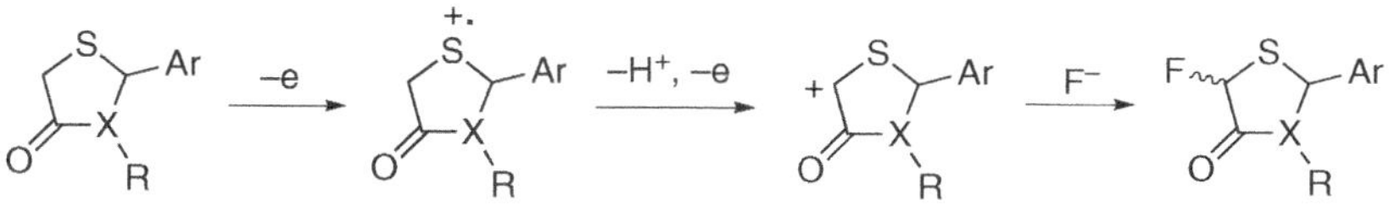

$\xrightarrow{-e}$ $\xrightarrow{-H^+,\ -e}$ $\xrightarrow{F^-}$

(X=N,O,S)

(4.10)

(b) Regioselectivity Controlled by Selective Adsorption of a Substrate to an Electrode due to its Dipole Moment As shown in Figure 4.5, tetrachloropicolinic acid is regioselectively dechlorinated by cathodic reduction. The high regioselectivity is attributed to the controlled orientation of the

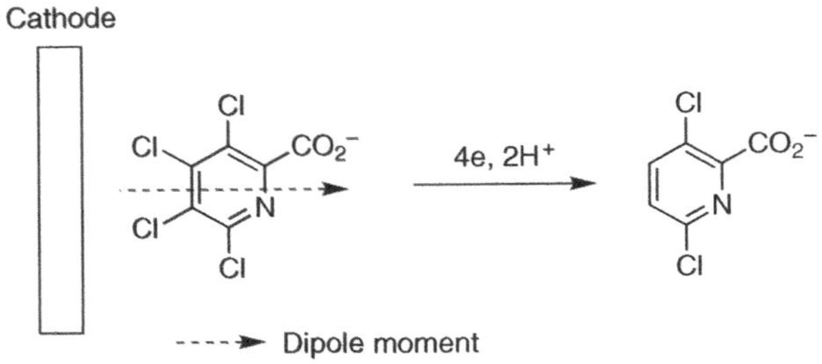

Figure 4.5 Regioselective cathodic dechlorination

substrate at the cathode surface owing to the dipole moment of the molecule [13].

4.3.2.4 *Stereoselectivity*

(a) Stereoselectivity Controlled by Selective Adsorption of Reactive Intermediate to Electrode As shown in Figure 4.6, the lithium salt of an unsaturated amine derivative adsorbs on the anode due to coulombic interactions. One-electron transfer then takes place to generate the corresponding aminyl radical intermediate, which undergoes subsequent intramolecular cyclization in an adsorbed manner or near the anode to result in predominant formation of the thermodynamically less stable *cis*-form product [14]. Interestingly, in this reaction the thermodynamically favoured *trans*-form product is not generated at all.

On the other hand, a similar reaction using a chemical oxidant such as $HgCl_2$ provides mainly the thermodynamically favourable *trans* product. Such high stereoselectivity in electrochemical reactions is mainly attributable to the adsorption effect. Thus, it can be stated that the electrode

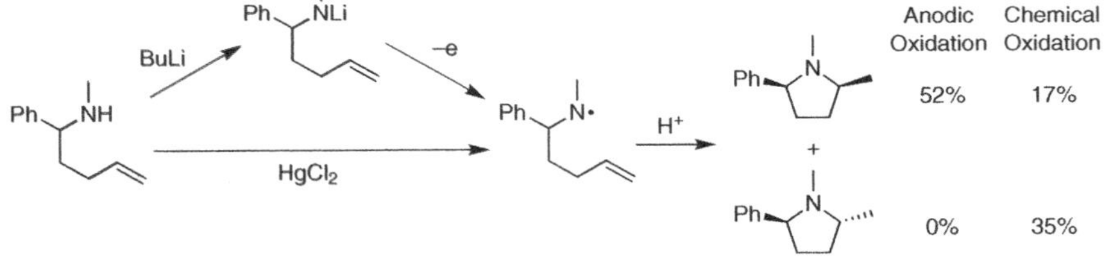

Figure 4.6 Stereoselective anodic cyclization

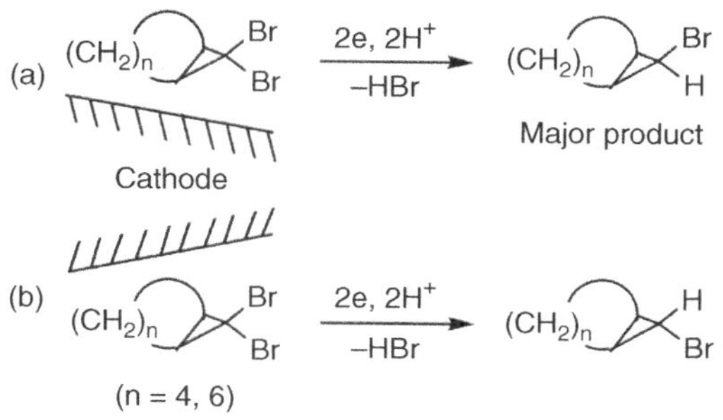

(a) $(CH_2)_n$ Br Br $\xrightarrow[-HBr]{2e,\ 2H^+}$ $(CH_2)_n$ Br H

Major product

Cathode

(b) $(CH_2)_n$ Br Br $\xrightarrow[-HBr]{2e,\ 2H^+}$ $(CH_2)_n$ H Br

(n = 4, 6)

Figure 4.7 Stereoselective reduction based on orientation of substrate on the cathode

contributes greatly to stereocontrol. In the cases of cathodic hydrogenation and anodic addition of cyclic olefins, *cis*-form products are preferentially formed. This is also attributable to addition of protons and nucleophiles to the anionic and cationic intermediates adsorbed on the electrode, respectively.

(b) Stereoselectivity Controlled by Steric Hindrance between Substrate and Electrode Simple stereocontrol is achieved by steric hindrance between substrate and electrode. As we can easily see in Figure 4.7, the orientation of bicyclic *gem*-dibromocyclopropane at the cathode surface as shown in (a) is favoured over that of the substrate shown in (b) owing to steric repulsion between the cathode and the 6- or 8-membered ring fused to cyclopropane. Therefore, the *exo*-bromide near the cathode is more easily reducible than the *endo*-bromide, and consequently the *endo*-bromide product is mainly formed [15].

4.3.2.5 Selectivity Depending on Electrode Materials

There are many examples of electrode materials that greatly affect product selectivity and stereoselectivity.

(a) Product Selectivity Acetone is reduced at a lead cathode in an acidic aqueous solution to give the corresponding alcohol, isopropyl alcohol, while the reduction with zinc and copper cathodes provides the corresponding alkane, propane [16].

(b) Anode Material Dependence of Kolbe Electrolysis It is well known that Kolbe electrolysis with a platinum anode generates a radical intermediate, while that with a carbon anode generates a cationic intermediate [1].

(c) Stereoselectivity The reduction of 2-methylcyclohexanone at zinc and copper cathodes in aqueous NaOH solution provides mainly the corresponding *trans*-alcohol, while that with a tin cathode provides *cis*-alcohol as a major product [17].

4.4 MOLECULAR ORBITALS AND ELECTRONS RELATED TO ELECTRON TRANSFER

The electrons involved in electrode electron transfer are not characteristic of electrolysis. From the viewpoint of molecular orbitals, the oxidation process can be explained by electron transfer from the highest occupied molecular orbital (HOMO) of a substrate molecule to the anode, while the reduction process is explained by electron transfer from the cathode to the lowest unoccupied molecular orbital (LUMO) of a substrate, as shown in Figure 4.8 [18].

In the case of a hydrocarbon, electron transfer takes place from unsaturated bonds and σ-bonds of strained compounds like cyclopropane or ionic species. However, ordinary saturated hydrocarbons rarely undergo redox reactions. On the other hand, even in the cases of saturated heteroatom compounds containing heteroatoms like N, S and O atoms, their oxidation is rather easy since electron transfer from lone paired electrons of a heteroatom readily takes place. Furthermore, depending on

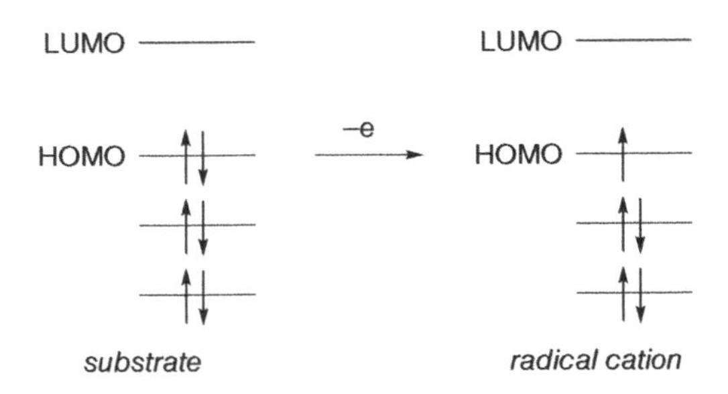

Figure 4.8 Molecular-orbital diagram for electron transfer

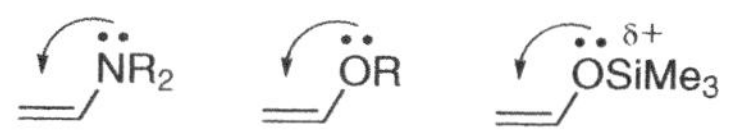

Figure 4.9 Easily oxidizable electron-rich olefins

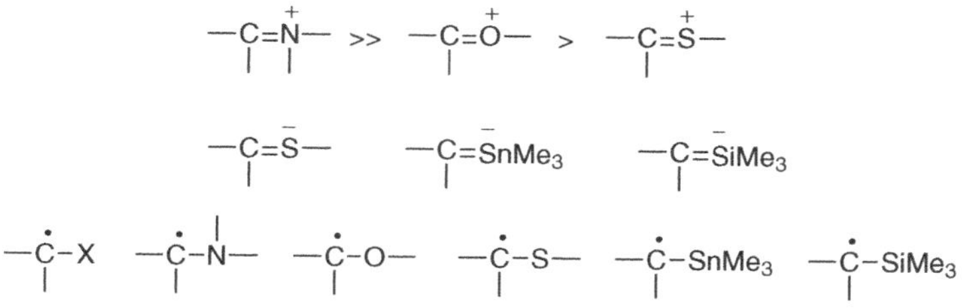

Figure 4.10 Stabilization effect of heteroatoms on electrogenerated reactive species

the heteroatom (Z), C–Z and Z–Z bonds can be reductively cleaved. Isolated olefins are generally difficult to oxidize, but olefins with a directly attached heteroatom, for instance enamines, enol ethers and silyl enol ethers (Figure 4.9), are electron-rich olefins, therefore their oxidation potentials decrease significantly and they are easily oxidized. Moreover, heteroatoms greatly contribute to the stability of electrochemically generated adjacent reactive intermediates, as shown in Figure 4.10. These effects are important characteristic properties of heteroatom compounds.

4.5 ELECTROAUXILIARIES

Electroauxiliaries are functional groups that facilitate electron transfer and control the reaction pathways of electrogenerated reactive species to provide the desired products selectively. Synthetic applications of electroauxiliaries in anodic oxidation are widespread while those for cathodic reduction are rare [18].

4.5.1 Electroauxiliaries Based on Molecular Orbital Interactions

The orbital interaction is highly effective in increasing the HOMO level. In principle, the interaction of the HOMO with a high-energy filled orbital increases its energy level. For instance, the energy level of a C–Si σ orbital

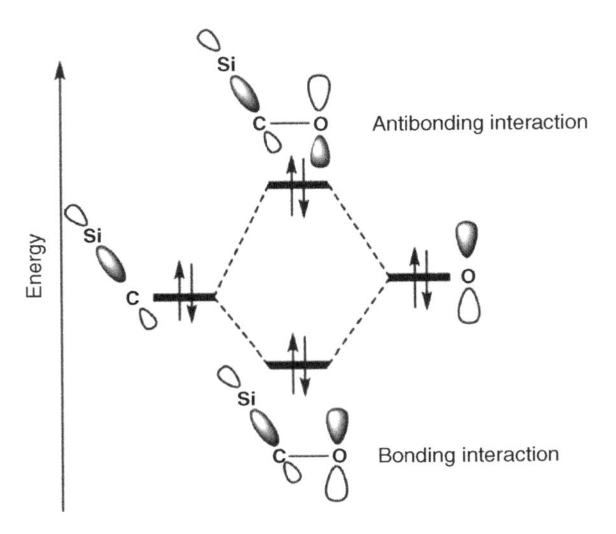

Figure 4.11 Interaction between the C–Si s orbital and the non-bonding 2p orbital of an oxygen atom to increase the HOMO level

is generally much higher than that of C–H and C–C σ orbitals therefore the C–Si σ orbital interacts effectively with a non-bonding *p* orbital of a heteroatom like N, O or S (σ–n interaction) if two orbitals align in the same plane (Figure 4.11) [18,19]. Accordingly, a silyl group at the α position activates a heteroatom compound towards anodic oxidation. One-electron oxidation gives the radical cation, in which the C–Si σ orbital interacts with a half-vacant *p* orbital of the heteroatom to stabilize the system. Such interaction also weakens the C–Si bond, and hence the C–Si bond is cleaved selectively. The resulting carbon radical undergoes further oxidation to give the carbocation, which is trapped by a nucleophile to give the desired product (Eq. 4.11). Thus, the silyl group not only activates substrates towards oxidation but also controls the reaction pathway [18,19].

R_3Si–CH(R)–Y $\xrightarrow{-e}$ $[R_3Si$–CH(R)–Y$]^{\bullet+}$ $\xrightarrow{-{}^+SiR_3}$ $R\dot{C}H$–Y $\xrightarrow{-e}$ $R\overset{+}{C}H$–Y $\xrightarrow{NuH}$ R–CH(Nu)–Y

Y = N, O, S, π-system

(4.11)

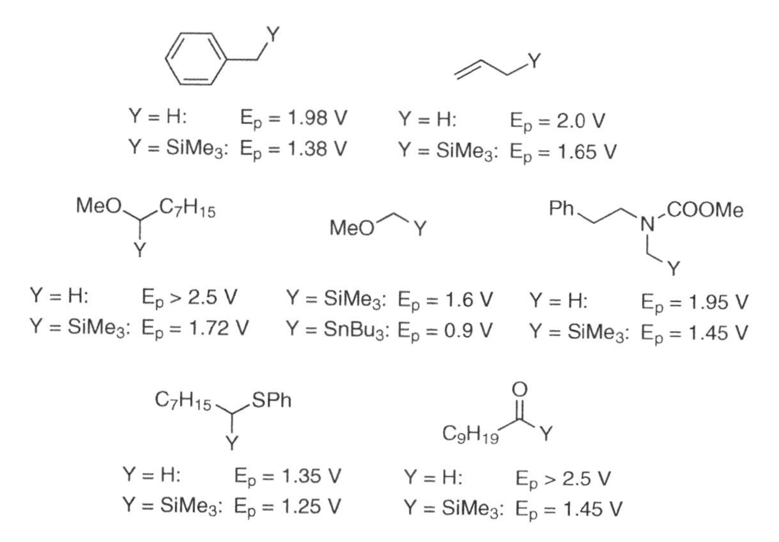

Figure 4.12 Example of decrease in oxidation potentials owing to molecular orbital interaction

In general, the electroauxiliary effect of a silyl group increases S < N < O, and the oxidation potential of a silyl compound with an α heteroatom decreases in the same order, as shown in Figure 4.12. In other words, the magnitude of the silicon effect for oxygen compounds is much greater than that for organo nitrogen and sulfur compounds. This is in accord with the better overlap of the non-bonding *p* orbital of oxygen with the C–Si σ orbital than that of the orbitals of nitrogen and sulfur [18,20,21].

The interaction of the C–Si σ orbital is also effective for raising the energy level of the adjacent π systems (σ–π interaction), and the C–Si bond is cleaved selectively. The silyl group therefore serves as an electroauxiliary for the oxidation of π systems. A stannyl group also serves as an electroauxiliary for the electrochemical oxidation of heteroatom compounds and π systems [18].

The anodic cyanation of *N*-benzyl-substituted piperizines usually provides a mixture of two regioisomeric products. However, the introduction of a silyl group as an electroauxiliary directs the reaction pathway to form a single product, as shown in Eq. 4.12 [22]. The silyl group also decreases the oxidation potential of the amine.

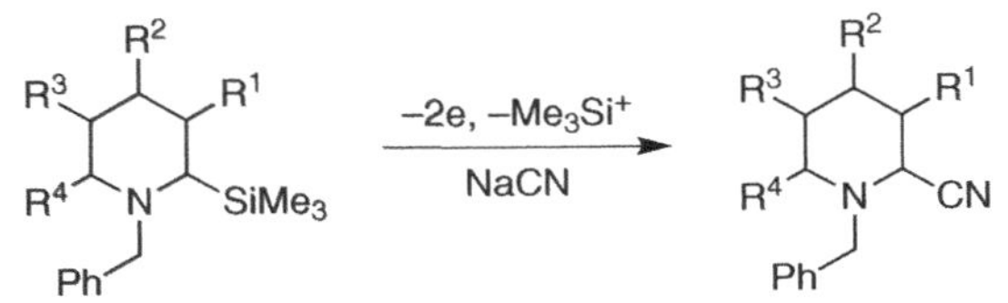

(4.12)

4.5.2 Electroauxiliaries Based on Readily Electron-Transferable Functional Groups

Easily oxidizable functional groups like the arylthio (ArS) group also work as electroauxiliaries, i.e. the introduction of an ArS group at the α position of ethers and carbamates decreases their oxidation potentials. Importantly, the oxidation of α-ArS substituted heteroatom compounds results in the selective cleavage of the C–S bond to generate a cation intermediate stabilized by the adjacent heteroatom. The resulting cation intermediate reacts with a nucleophile to provide the product selectively. The use of an ArS group as an electroauxiliary expands the scope of nucleophiles, which enables *in situ* use of carbon nucleophiles like allylsilane, as shown in Eqs. 4.13 and 4.14, even if such nucleophiles are easily oxidized [23]. In these cases, owing to the ArS group, selective C–S bond cleavage takes place and an allyl group is introduced to the α or benzylic carbon selectively.

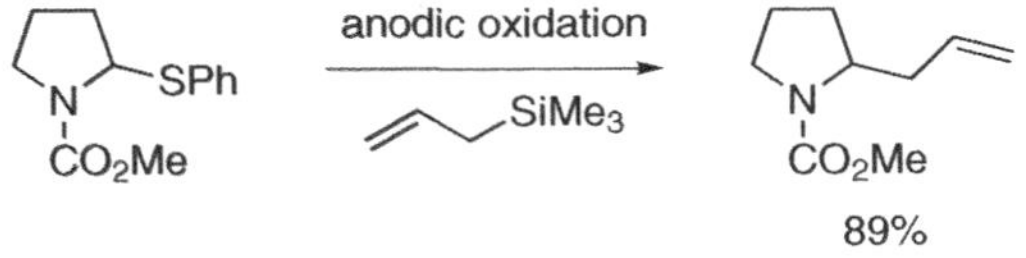

(4.13)

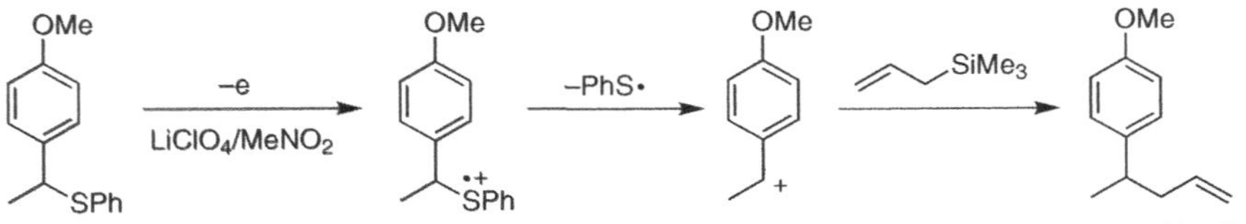

(4.14)

On the other hand, it is known that the electroauxiliary moiety and reactive site are different in the case of cathodic reduction. For instance, as shown in Eq. 4.15, the first electron transfer takes place at the easily electron acceptable carbonyl group, and then adjacent carbon–heteroatom bond cleavage takes place selectively [24].

$$\mathrm{RCOCH_2Y} \xrightarrow{e} [\,\mathrm{R\dot{C}(O^-)CH_2Y}\,] \xrightarrow[-\mathrm{Y}]{e,\ \mathrm{H^+}} \mathrm{RCOCH_3} \qquad (4.15)$$

(Y=OR', SR', NR'R'')

Based on this principle, selective C–O bond cleavage at the benzylic and allylic positions can be easily achieved, as shown in Eq. 4.16.

$$\text{2-(ROCOCH}_2\text{)-anthraquinone} \xrightarrow{2e,\ \mathrm{H^+}} \text{2-(CH}_3\text{)-anthraquinone} \qquad (4.16)$$

Although the C–F bond is not so easily cleaved by reduction owing to its larger bond energy compared to that of the C–H bond, fluorine attached to a benzylic position or α to carbonyl and imino groups is readily removed by cathodic reduction. This is also quite similar to the cases shown in Eqs. 4.15 and 4.16. As shown in Eq. 4.17, electron transfer first takes place at the aromatic ring or carbonyl and imino groups followed by β-elimination of the fluoride ion [25]. When the cathodic reaction shown in Eq. 4.17 is carried out in the absence of Me_3SiCl, further cathodic reduction takes place to result in successive elimination of fluorine atoms, providing complicated products.

$$\mathrm{F_3C{-}C({=}X){-}Ar} \xrightarrow[\mathrm{Me_3SiCl}]{2e,\ -\mathrm{F^-}} \mathrm{F_2C{=}C(XSiMe_3){-}Ar} \qquad (4.17)$$

(X = O, NR)

4.5.3 Electroauxiliaries Based on Intermolecular Coordination Effects

This effect is quite important from an electron transfer aspect. The electron transfer reaction in solution is generally facilitated by the stabilization of the resulting radical ion and ionic intermediates by the coordination of solvent molecules or counter ions in the solution. For instance,

(Z = O, NR)

$R{-}C{\equiv}N \xrightarrow{H^+} R{-}C{\equiv}\overset{+}{N}H$

$ArNO_2$ (R) $\xrightarrow{H^+}$ Ar(R)–$\overset{+}{N}$(OH)=O

Figure 4.13 Anodic shift of reduction potential owing to protonation

the reduction potential of the metal cation usually becomes more negative in solvents with large donor numbers because the positive nature of the metal cation is decreased by solvation in such solvents.

It is well known that the reduction potentials of ketones, imines, cyano and nitro compounds shift in the positive direction by proton addition. The polarography of the reduction of nitro compounds has been intensively studied. When the pH value of the solution is made more acidic by one pH unit, the reduction potential becomes more positive by about 58 mV. The positive shift of the reduction potential is due to the protonation of unsaturated functional groups resulting in a positively charged form, as shown in Figure 4.13.

Furthermore, the reduction potentials of ketones often shift to positive in the presence of Lewis acids. This is due to the coordination of the Lewis acid to the oxygen atom of the carbonyl group resulting in a decrease in the electron density of the carbonyl group or coordination of the Lewis acid to the anion radical intermediate generated by one-electron reduction of the carbonyl group resulting in stabilization of the intermediate, as shown in Figure 4.14.

(L = Lewis acid)

RO–H····O⁻–(anthracene)–O⁻····H–OR

Figure 4.14 Anodic shift of reduction potential owing to coordination with Lewis acid and hydrogen bonding

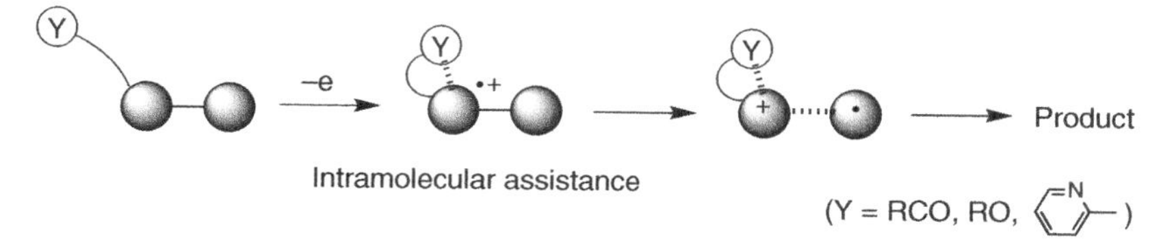

Figure 4.15 Cathodic shift of oxidation potential owing to intramolecular coordination with a suitable functional group (Y)

It is also known that the second reduction potentials of diketones like anthraquinone and dinitrobenzenes shift greatly in the positive direction in the presence of alcohols. This is due to the formation of hydrogen bonds between dianion intermediates generated by two-electron reduction and the alcohols.

4.5.4 Electroauxiliaries Based on Intramolecular Coordination Effects

If a substrate molecule has a specific coordinating site such as a functional group or heteroatom to stabilize the electrogenerated ionic intermediate, the electron transfer is enhanced by intramolecular coordination, as shown in Figure 4.15. As such a functional group, pyridyl, carbonyl and ether groups are effective, and they decrease the oxidation potential appreciably. Such coordination would facilitate subsequent chemical steps like bond fission. When bond cleavage accompanies electron transfer, as shown in Figure 4.15, the oxidation potential decreases significantly. This is quite a new concept and methodology for electron transfer control [26]. Furthermore, it has been demonstrated that a combination of the orbital interaction and intramolecular coordination is highly effective for controlling the oxidation potential as well as the reaction pathway.

4.6 REACTION PATTERN OF ORGANIC ELECTRODE REACTIONS

Although electrode reactions are limited to oxidation and reduction, various follow-up chemical reactions often take place after organic electron transfer. Organic electrode reactions are therefore quite different from those of inorganic compounds.

From a synthetic viewpoint, organic electrode reactions are classified into various reaction types, and examples of cathodic reduction and anodic oxidation are given here.

4.6.1 Transformation Type of Functional Group

Cathodic Reduction

$$R{-}NO_2 \xrightarrow[-H_2O]{2e,\,2H^+} R{-}NO \xrightarrow{2e,\,+2H^+} R{-}NHOH \xrightarrow[-H_2O]{2e,\,2H^+} R{-}NH_2 \quad (4.18)$$

$$RR'C{=}O \xrightarrow{2e,\,2H^+} RR'CH{-}OH; \qquad RR'C{=}O \xrightarrow[-H_2O]{4e,\,4H^+} RR'CH_2 \quad (4.19)$$

$$Ar{-}COOR\,(H) \xrightarrow{4e,\,4H^+} Ar{-}CH_2OH + ROH \quad (4.20)$$

$$R{-}CONR'_2 \xrightarrow{4e,\,4H^+} R{-}CH_2NR'_2 + H_2O \quad (4.21)$$

$$(R{-}CO)_2NR' \xrightarrow{8e,\,8H^+} (R{-}CH_2)_2NR' + 2H_2O \quad (4.22)$$

$$Ar{-}N_2^+Cl^- \xrightarrow{4e,\,4H^+} Ar{-}NH{-}NH_2 + HCl \quad (4.23)$$

$$R_2C{=}N{-}R' \xrightarrow{2e,\,2H^+} R_2CH{-}NH{-}R' \quad (4.24)$$

$$R_2C{=}N{-}OH \xrightarrow{4e,\,4H^+} R_2CH{-}NH_2 + H_2O \quad (4.25)$$

$$R{-}C{\equiv}N \xrightarrow{2e,\ 2H^+} [R{-}CH{=}NH] \xrightarrow{2e,\ 2H^+} R{-}CH_2NH_2 \quad (4.26)$$

$$Ar{-}SO_2H \xrightarrow{4e,\ 4H^+} Ar{-}SH + 2H_2O \quad (4.27)$$

$$-C{\equiv}C- \xrightarrow{2e,\ 2H^+} -CH{=}CH- \xrightarrow{2e,\ 2H^+} -CH_2{-}CH_2- \quad (4.28)$$

$$R{-}C_6H_5 \xrightarrow{2e,\ 2H^+} R{-}\text{cyclohexadiene} \xrightarrow{2e,\ 2H^+} R{-}\text{cyclohexene} \xrightarrow{2e,\ 2H^+} R{-}\text{cyclohexane} \quad (4.29)$$

Anodic Oxidation

$$R{-}CH_2OH \xrightarrow{-2e,\ -2H^+} R{-}CHO \xrightarrow{-2e,\ 2OH^-} R{-}COOH + H_2O \quad (4.30)$$

$$RR'CHOH \xrightarrow{-2e,\ -2H^+} RR'C{=}O \quad (4.31)$$

$$Ar{-}CH_3 \xrightarrow[+2OH^-]{-4e,\ -2H^+} Ar{-}CHO \xrightarrow[+OH^-]{2e,\ -H^2} Ar{-}COOH \quad (4.32)$$

$$Ar{-}NH{-}NH{-}Ar \xrightarrow{-2e,\ -2H^+} Ar{-}N{=}N{-}Ar \xrightarrow[+2OH^-]{-2e,\ -2H^+} Ar{-}N{=}N(\rightarrow O){-}Ar \quad (4.33)$$

$$R{-}S{-}R \xrightarrow[+OH^-]{-2e,\ -H^+} R{-}S({=}O){-}R \xrightarrow[+OH^-]{-2e,\ -H^+} R{-}S({=}O)_2{-}R \quad (4.34)$$

$$R{-}CH_2NH_2 \xrightarrow{-4e,\ -4H^+} R{-}C{\equiv}N \quad (4.35)$$

$$R{-}SH \xrightarrow{-e,\ -H^+} 1/2\ R{-}S{-}S{-}R \quad (4.36)$$

Equations 4.18–4.36 are shown using stoichiometric formula including elementary reaction steps, but only final products are shown and elementary reaction steps are omitted in the following examples. It should also be noted that the reagents shown under the arrows react with electrochemically generated intermediates, but they are not subjected to direct electron transfer reactions.

4.6.2 Addition Type

Cathodic Reduction

$$R{-}C(=O){-}R' \xrightarrow[CO_2]{2e,\ 2H^+} R{-}C(COOH)(OH){-}R' \quad (4.37)$$

$$R{-}C(=O){-}R' \xrightarrow[\text{alkene}]{2e,\ 2H^+} R'{-}C(R)(OH){-}C{-}CH \quad (4.38)$$

$$H_2C{=}CH{-}CH{=}CH_2 \xrightarrow[2CO_2]{2e,\ 2H^+} HOOC{-}CH_2{-}CH{=}CH{-}CH_2{-}COOH \quad (4.39)$$

$$R_2CX_2\ \text{(X: halogen)} \xrightarrow[\text{alkene}]{2e,\ -2X^-} \text{cyclopropane}(R)(R) \quad (4.40)$$

Anodic Oxidation

$$\text{furan} \xrightarrow[2ROH]{-2e,\ -2H^+} \text{2,5-di(RO)-2,5-dihydrofuran (RO, OR)} \quad (4.41)$$

$$2R{-}COOH + \text{alkene} \xrightarrow{-2e,\ -2H^+} R{-}COO{-}C{-}C{-}OOC{-}R \quad (4.42)$$

$$I^- + \text{>C=C<} \xrightarrow[H_2O/CH_3CN]{-2e} I-\overset{|}{\underset{|}{C}}-\overset{|}{\underset{|}{C}}-NHCOCH_3 \quad (4.43)$$

4.6.3 Insertion Type

Cathodic Reduction

$$\underset{(Y = ROCO,\ ArSO_2)}{Y-NCl_2} \xrightarrow[\text{1,4-dioxane}]{2e,\ -2Cl^-} \text{2-(YNH)-1,4-dioxane} \quad (4.44)$$

Anodic Oxidation

$$R_2N-NH_2 \xrightarrow[R'-NH_2]{-2e,\ -2H^+} R_2N-NH-NH-R' \quad (4.45)$$

4.6.4 Substitution Type

Cathodic Reduction

$$\underset{(X = I,\ Br,\ Cl)}{R-X} \xrightarrow[H^+]{2e,\ -X^-} R-H \quad (4.46)$$

$$-\overset{|}{\underset{|}{C}}-H \xrightarrow[Y^+]{e,\ -1/2H_2} -\overset{|}{\underset{|}{C}}-Y \quad (4.47)$$

Anodic Oxidation

$$R-H \xrightarrow[X^-]{-2e,-H^+} R-X \quad (X=I,\ Br,\ Cl,\ F) \quad (4.48)$$

$$R-C_6H_5 + Y^- \xrightarrow{-2e,\ -H^+} R-C_6H_4-Y \quad (4.49)$$

$(R = NH_2,\ OR')$ $(Y = RCOO,\ NCS,\ CN,\ RO,\ NCO,\ \text{halogen})$

$$R-C(=O)-N(R')-CH_3 \xrightarrow[R''COOH]{-2e,\ -2H^+} R-C(=O)-N(R')-CH_2-OOCR'' \quad (4.50)$$

$$Ar-N(R)-CH_3 \begin{cases} \xrightarrow[R'O^-]{-2e,\ -H^+} Ar-N(R)-CH_2-OR' \\ \xrightarrow[R'COO^-]{-2e,\ -H^+} Ar-N(R)-CH_2-OOCR' \\ \xrightarrow[CN^-]{-2e,\ -H^+} Ar-N(R)-CH_2-CN \end{cases} \quad (4.51)$$

4.6.5 Substitutive Exchange Type

Cathodic Reduction

$$X-C_6H_4-NO_2 \xrightarrow[OH^-]{6e,\ 6H^+} HO-C_6H_4-NH_2 + X^- \quad (4.52)$$

(X = halogen, OR)

In Eq. 4.52, the transformation type of functional group also occurs.

Anodic Oxidation

$$R-COO^- \xrightarrow[Y^-]{-2e} R-Y + CO_2 \quad (4.53)$$

$$(Y = R'COO, R'O, NO_3, HO)$$

$$\underset{(X = I, Br)}{R-X} \xrightarrow[H_2O/CH_3CN]{-e,\ -1/2X_2} R-NHC(=O)-CH_3 \quad (4.54)$$

4.6.6 Elimination Type

Cathodic Reduction

$$-C(X)-C(Y)- \xrightarrow{2e} -C=C- + X^- + Y^- \ \text{(1,2-elimination)} \quad (4.55)$$

(X, Y = halogen, RCOO, RSO_3, RS, HS, HO; X = Y = O–CO–O)

$$R-CH(X)-CH_2-CH(X)-R' \xrightarrow{2e,\ -2X^-} R\text{-}\triangle\text{-}R' \quad \text{(1,3-elimination)} \tag{4.56}$$

Anodic Oxidation

$$-\overset{|}{\underset{HOOC}{C}}-\overset{|}{\underset{COOH}{C}}- \xrightarrow{-2e,\ -2H^+} -\overset{|}{C}=\overset{|}{C}- + 2CO_2 \quad \text{(1,2-elimination)} \tag{4.57}$$

4.6.7 Dimerization Type

Cathodic Reduction

$$RR'C=O \xrightarrow{e,\ H^+} 1/2\ R'-\overset{R}{\underset{OH}{C}}-\overset{R}{\underset{OH}{C}}-R' \tag{4.58}$$

$$RR'C=NR'' \xrightarrow{e,\ H^+} 1/2\ R'-\overset{R}{\underset{R''HN}{C}}-\overset{R}{\underset{NHR''}{C}}-R' \tag{4.59}$$

$$>C=C<_{Y} \xrightarrow{e,\ H^+} 1/2\ Y-CH-\overset{|}{\underset{|}{C}}-\overset{|}{\underset{|}{C}}-CH-Y \tag{4.60}$$

(Y = CN, COOR, $CONR_2$, RCO)

$$R-X \xrightarrow{e} 1/2R-R+X^- \tag{4.61}$$

(X=halogen)

$$Ar-NO_2 \xrightarrow[-3/2H_2O]{3e,\ 3H^+} 1/2\ Ar-N=\underset{\downarrow \atop O}{N}-Ar \xrightarrow[-1/2H_2O]{e,\ H^+} 1/2\ Ar-N=N-Ar \xrightarrow[-1/2H_2O]{e,\ H^+} 1/2\ Ar-NH-NH-Ar \tag{4.62}$$

Anodic Oxidation

$$R{-}COO^- \xrightarrow{-2e} 1/2R{-}R + CO_2 \quad (4.63)$$

$$R{-}C_6H_5 \xrightarrow{-e,\,-H^+} 1/2\ R{-}C_6H_4{-}C_6H_4{-}R \quad (4.64)$$

$$>C{=}C< + ROH \xrightarrow{-e,\,-H^+} 1/2\ RO{-}\overset{|}{\underset{|}{C}}{-}\overset{|}{\underset{|}{C}}{-}\overset{|}{\underset{|}{C}}{-}\overset{|}{\underset{|}{C}}{-}OR \quad (4.65)$$

$$(R_2N)_2C{=}CH_2 \xrightarrow{-e,\,-H^+} 1/2\ (R_2N)_2C{=}CH{-}CH{=}C(NR_2)_2 \quad (4.66)$$

4.6.8 Crossed Dimerization

$$RR'C{=}O + >C{=}C<_{Y} \xrightarrow{2e,\,2H^+} R'{-}\underset{OH}{\overset{R}{C}}{-}\overset{|}{\underset{|}{C}}{-}\overset{H}{\underset{|}{C}}{-}Y \quad (4.67)$$

(Y = COOR, CN, $CONH_2$)

In Eq. 4.67, the ratio of crossed dimerization to non-crossed dimerization can be increased, but this is not so easy in the case of Eq. 4.68.

$$R{-}COOH + R'{-}COOH \xrightarrow[-2CO_2]{-2e,-2H^+} R{-}R' \quad (4.68)$$

4.6.9 Cyclization Type

$$R{-}\underset{\parallel}{\overset{}{C}}{-}R'{-}CH{=}CH{-}R'' \;(C{=}O) \xrightarrow{2e,\,2H^+} R(HO)\overset{}{C}{<}(R'){>}CH{-}CH_2{-}R'' \quad (4.69)$$

$$(CH_2)_m{<}^{CH_2}_{NH}{>}(CH_2)_n \xrightarrow{-2e,\,-2H^+} (CH_2)_m{<}^{CH}_{N}{>}(CH_2)_n \ \text{(bicyclic, CH–N bridge)} \quad (4.70)$$

4.6.10 Polymorphism Formation Type

$$CS_2 \xrightarrow{e} 1/6\ (^{-}S)_2C_2S_2C{=}CS_2C_2(S^{-})_2 \tag{4.71}$$

4.6.11 Polymerization Type

$$CH_2{=}CH{-}Y \xrightarrow[\text{(catalytic amount)}]{e} 1/n \left[CH_2{-}\underset{Y}{CH} \right]_n \tag{4.72}$$

(Y = COOR, CN, $CONH_2$, Ph, AcO, alkyl)

The reaction shown in Eq. 4.72 is initiated by cathodic reduction of activated olefin and a catalytic amount of electricity is enough to facilitate the polymerization in a chain reaction mechanism, while the following reaction is anodic oxidative polymerization consuming stoichiometric amount of electricity.

$$C_4H_4Y \xrightarrow{-2ne,\ -2nH^+} 1/n \left[C_4H_2Y \right]_n \tag{4.73}$$

(Y = NH, S)

4.6.12 Cleavage Type

$$R{-}S{-}S{-}R \xrightarrow{2e,\ 2H^+} 2R{-}SH \tag{4.74}$$

$$-\underset{X}{\overset{|}{C}}-\underset{Y}{\overset{|}{C}}- \xrightarrow[-HX,\ -HY]{-2e,\ 2OH^-} 2\ {>}C{=}O \tag{4.75}$$

(X, Y = NR_2, OR)

4.6.13 Metalation Type

Metal used as an electrode is incorporated in the product.

$$RR'C{=}O + 1/2\ M\ (\text{Cathode}) \xrightarrow[-OH^-]{3e,\ 2H^+} 1/2\ RR'CH{-}M{-}CHRR' \tag{4.76}$$

(M = Hg, Pb)

$$\mathrm{RMgX} + 1/4\mathrm{Pb\ (Anode)} \xrightarrow[-[\mathrm{MgX}]^+]{-e} 1/4\mathrm{R_4Pb} \quad (X = \text{halogen}) \tag{4.77}$$

4.6.14 Asymmetric Synthesis Type

$$\mathrm{RR'C{=}Z} \xrightarrow[(X = O,\ NR'')]{2e,\ 2H^+} \mathrm{R'{-}C^*(R)(H){-}ZH} \tag{4.78}$$

$$\mathrm{R{-}S{-}R'} \xrightarrow[-H^+]{-2e,\ OH^-} \mathrm{R{-}S^*({=}O){-}R'} \tag{4.79}$$

$$\mathrm{RR'C{=}C\langle} \xrightarrow{2e,\ 2H^+} \mathrm{R'{-}C^*(R)(H){-}CH\langle} \tag{4.80}$$

As a chiral source, chiral supporting salts, chiral solvents, chiral adsorbants and chiral modified electrodes are used.

4.7 ELECTROCHEMICALLY GENERATED REACTIVE SPECIES

Electrochemically generated reactive species often undergo follow-up reactions in an adsorbed state or near the electrode surface to provide products. Since these reactive species are affected by the electrode itself and a strong electric field (approximately 10^7–10^8 V cm^{-1}), their reactivity and behaviour are often quite different from the same reactive species generated by other methods. Useful synthetic reactions can be developed utilizing such unique electrogenerated reactive species. A variety of ionic and radical species are generated by electrolysis. Electrochemically generated reactive species are classified into carbon species and heteroatom ones, and their generation methods are explained below.

4.7.1 Carbon Species

4.7.1.1 Anodically Generated Carbon Species

Oxidation of Carboxylic Acid: Anodic oxidative decarboxylation of carboxylic acid generates an alkyl radical and/or alkyl cation [27]. A carboxylic acid with a straight alkyl chain is oxidized at the Pt anode in weak acidic solution to generate mainly the corresponding alkyl radical. On the other hand, anodic oxidation of carboxylic acid with an α-branched alkyl chain at a carbon anode in neutral or alkaline solution generates mainly the corresponding alkyl cation.

$$R{-}COOH \xrightarrow{-e,\,-H^+} R{-}COO^{\bullet} \xrightarrow{-CO_2} R^{\bullet} \xrightarrow{-e} R^+ \quad (4.81)$$

Oxidation of Carbanion and Active Hydrogen Compounds: The oxidation potentials of carbanions are extremely low and their one-electron oxidation generates radicals.

$$R^- \xrightarrow{-e} R^{\bullet} \quad (4.82)$$

$$X{-}CH_2{-}Y \xrightarrow{-e,\,-H^+} X{-}\dot{C}H{-}Y \xrightarrow{-e} X{-}\overset{+}{C}H{-}Y$$

$$(X, Y{=}CN, COR, COOR, NO_2) \quad (4.83)$$

Oxidation of Alkyl Halides: Alkyl iodides are more easily reduced and oxidized compared to the corresponding bromides. Their one-electron oxidation generates a radical cation intermediate followed by elimination of the halogen molecule to generate the alkyl cation.

$$R{-}X \xrightarrow{-e} [R{-}X]^{\bullet+} \xrightarrow{-1/2X_2} R^+$$

$$(X{=}Br, I) \quad (4.84)$$

Oxidation of Olefins, Ketones, Imines, and Strain and Cage Compounds: The oxidation potential of ketone is generally high, but even aliphatic ketone is oxidizable. Although alkane is difficult to oxidize, strain cyclopropanes and cage-type adamantanes are relatively easy to oxidize. One-electron oxidation of those compounds generates radical cations.

$$\gt C{=}Z \xrightarrow{-e} \gt \overset{+}{C}{-}\dot{Z} \quad (Z = C, O, N) \tag{4.85}$$

$$-\overset{|}{C}{-}C{-}C{-} \xrightarrow{-e} \gt\overset{+}{C}\;\;\dot{C}\lt \text{ (bridged by } \gt C \lt) \tag{4.86}$$

Oxidation of Aromatic Compounds: Initially, one-electron transfer from the aromatic ring occurs to generate a radical cation on the ring. When an alkyl chain exists on the aromatic ring, elimination of the α-proton from the alkyl chain generates the corresponding benzyl radical, which undergoes further one-electron oxidation to generate benzyl cation. On the other hand, in the case of difficult elimination of the α-proton or aromatics devoid of α-proton, the aromatic radical cation is attacked by nucleophiles, resulting in ring substitution with the nucleophiles. In the absence of nucleophiles, homo-coupling of radical cations or radical intermediates takes place.

$$Ar{-}CH_2{-}R \xrightarrow{-e} \overset{\bullet+}{Ar}{-}CH_2{-}R \xrightarrow{-H^+} Ar{-}\dot{C}H{-}R \xrightarrow{-e} Ar{-}\overset{+}{C}H{-}R \tag{4.87}$$

$$Ar{-}OR \xrightarrow{-e} \overset{\bullet+}{Ar}{-}OR \xrightarrow{-H^+} \dot{Ar}{-}OR \xrightarrow{-e} \overset{+}{Ar}{-}OR \tag{4.88}$$

4.7.1.2 *Cathodically Generated Carbon Species*

Reduction of Alkyl Halides: The ease of reduction of alkyl halides is related to the bond energy of C–X, and hence iodide compounds are the most easily reduced, while chloride compounds are the most difficult to reduce. One-electron and two-electron reduction of alkyl halides generate alkyl radicals and alkyl anions, respectively.

$$R{-}X \xrightarrow{e,\,-X^-} R^\bullet \xrightarrow{e} R^- \quad (X{=}Cl, Br, I) \tag{4.89}$$

Reduction of Ketone and Imine: In the case of cathodic reduction of ketone and imine in aqueous solution, the generated reactive species as well as the reaction mechanism are changed by the pH of the solution. In an acidic solution, the oxygen atom of the ketone and the nitrogen atom of the imine are protonated, therefore their reduction potentials shifts to the positive side, and their one-electron reduction generates neutral radicals. In contrast, in an alkaline solution the protonation of ketone and imine does not occur due to low proton concentration. In this case, the radical anion is generated first, and then the dianion is formed.

$$\mathrm{>C{=}Z} \xrightarrow{e} \mathrm{>\dot{C}{-}Z^{-}} \begin{cases} \xrightarrow{H^+} \mathrm{>\dot{C}{-}ZH} \xrightarrow{e} \\ \xrightarrow{e} \mathrm{>\bar{C}{-}Z^{-}} \xrightarrow{H^+} \end{cases} \longrightarrow \mathrm{>\bar{C}{-}ZH} \qquad (4.90)$$

(Z = O, N)

Reduction of Activated Olefin and Conjugated Olefin: Activated olefin is readily reduced because the electron-withdrawing group attached to the double bond decreases the electron density of the double bond. Isolated olefin is not so easy to reduce, but conjugated olefin is reducible. One-electron reduction of the olefin generates a radical anion.

$$\mathrm{>C{=}C<^{Y}} \xrightarrow{e} \mathrm{-\dot{C}{-}\bar{C}{-}Y} \qquad (4.91)$$

(Y = CN, COR, COOR, >C=C<)

Reduction of Active Hydrogen Compounds: One-electron reduction of an active hydrogen compound generates the corresponding anion, eliminating the active hydrogen atom as hydrogen gas.

$$\mathrm{X{-}CH_2{-}Y} \xrightarrow{e,-1/2H_2} \mathrm{X{-}\bar{C}H{-}Y} \qquad (4.92)$$

(X, Y=CN, COR, COOR)

Reduction of *gem*- and *vic*-Dihalogeno Compounds: Two-electron reduction of these compounds generates carbene and benzyne.

$$\mathrm{>CX_2} \xrightarrow{2e,\ -2X^-} \mathrm{>C:} \qquad (4.93)$$

(X = Cl, Br, I) (Carbene)

$$\text{C}_6\text{H}_4\text{X}_2 \xrightarrow{2e,\ -2X^-} \text{C}_6\text{H}_4 \tag{4.94}$$

(X = Cl, Br, I) (Benzyne)

4.7.2 Heteroatom Species

Different from hydrocarbons, heteroatom compounds are oxidizable since they have lone paired electrons on the heteroatom, from which electron transfer occurs. In general, ease of oxidation has the order N > S > O, and a variety of nitrogen active species are generated electrochemically [28].

4.7.2.1 *Nitrogen Species*

One-electron oxidation of amine generates a radical cation at the nitrogen atom. The radical cation of aromatic amine is relatively stable, while that of aliphatic amine is so unstable that the α-proton is immediately eliminated, and then the active site shifts to the α-carbon. In the case of ordinary aliphatic amines, iminium ions are so unstable that cleavage of the C–N bond occurs predominantly. On the other hand, iminium ions of aromatic amines, carbamates and amides are stable, therefore nucleophilic substitution reactions like methoxylation, acetoxylation and cyanation occur efficiently at the α-position to the nitrogen atom, as shown in Eq. 4.95 [6]. These anodic substitutions are one of the characteristic electrolytic reactions and they are useful for construction of a C–C bond at the adjacent position to the nitrogen atom. Anodic substitution at the adjacent position to the nitrogen atom of imines is also possible, as shown in Eq. 4.96 [29].

$$\text{>N-CH}_2\text{-} \xrightarrow{-e} \text{>N}^{\bullet+}\text{-CH}_2\text{-} \xrightarrow{-H^+} \text{>N-}\dot{\text{C}}\text{H-}$$

$$\downarrow -e$$

$$\text{>NH} + \text{-CHO} \xleftarrow[-H^+]{H_2O} \left[\text{>N}^+\text{=CH-} \longleftrightarrow \text{>N-}\overset{+}{\text{C}}\text{H-}\right] \xrightarrow{Nu^-} \text{>N-CH(Nu)-}$$

(Nu = MeO, AcO, CN, etc.)

(4.95)

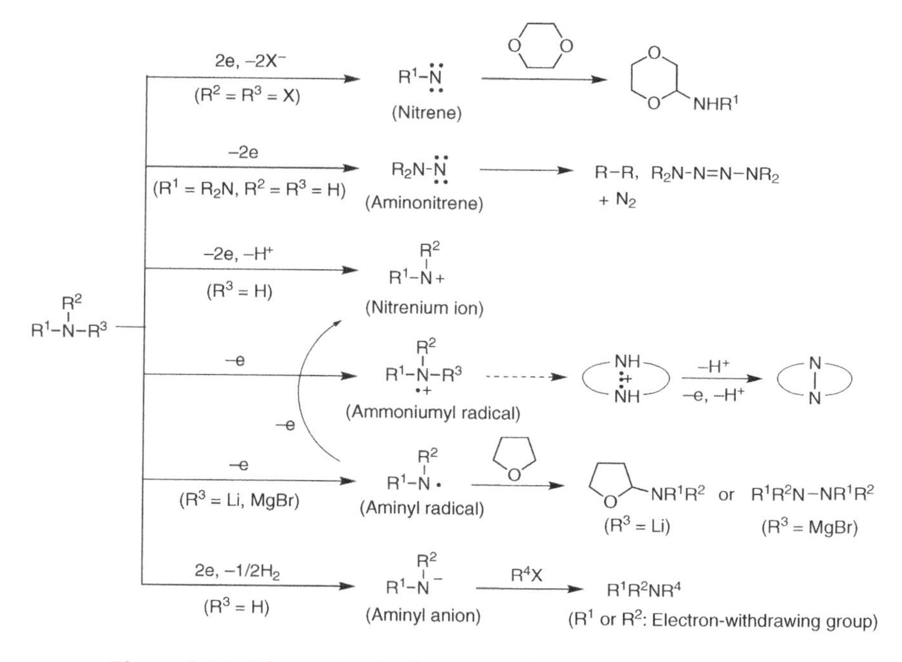

Figure 4.16 Electrochemically generated reactive nitrogen species

$$\xrightarrow{-2e,\ -H^+} \left[\ \cdots\ \right] \xrightarrow{Nu^-} \qquad (4.96)$$

(Nu = MeO, AcO, F)

Depending on the molecular structures of amines, the active site is retained at the original nitrogen, resulting in the generation of various reactive nitrogen species, and eventually reactions occur at the nitrogen atom, as shown in Figure 4.16 [28]. Two-electron reduction of *N*,*N*-dihalo compounds generates the corresponding nitrenes [30], while two-electron oxidation of aziridine and medium-sized cyclic amines followed by deprotonation of NH moiety generates nitrenium ions [31]. Anodic oxidation of hydrazine derivatives generates amino nirenes [32]. In particular, anodic oxidation of *N*-phthalimide in the presence of olefin provides the corresponding aziridine derivatives in high yields, as shown in Figure 4.16 [33]. The reaction proceeds via anodically generated aminonitrene.

$$\text{Phthalimide-N-NH}_2 + \text{R}^1\text{CH=CHR}^2 \xrightarrow[\text{Et}_3\text{NHOAc, r.t.}]{-2e,\ -H^+} \text{Phthalimide-N-N(aziridine-R}^1\text{,R}^2) \tag{4.97}$$

Furthermore, anodic oxidation of bicyclic amines and aliphatic amines devoid of α-protons generates aminyl radicals efficiently. In the former case, anodic elimination of α-proton cannot occur due to Bredt's rule.

$$\text{7-azabicyclo[2.2.1]heptane (NH)} \xrightarrow{-e,\ -H^+} \text{N}\cdot \tag{4.98}$$

$$t\text{-BuNH}_2 \xrightarrow{-e, -H^+} t\text{-Bu}\dot{\text{N}}\text{H} \tag{4.99}$$

4.7.2.2 Oxygen Species

Oxidation of Alcohol and Carboxylic Acid: Radical species are mainly generated from these compounds, but cationic species could also be generated.

$$\text{RO}^- \xrightarrow{-e} \text{RO}^\bullet \xleftarrow{-e, -H^+} \text{ROH} \tag{4.100}$$

$$\left.\begin{array}{l}\text{R-COO}^- \xrightarrow{-e} \\ \text{R-COOH} \xrightarrow{-e,\ -H^+}\end{array}\right\} \longrightarrow \text{R-COO}\cdot \xrightarrow{-e} \text{R-COO}^+ \tag{4.101}$$

Oxidation of Ether: Similarly to the case of organonitrogen compounds, anodic oxidation of ethers generates a cation α to the oxygen atom, which undergoes nucleophilic substitution. However, the scope of its application is limited because of the high oxidation potentials of ethers.

$$\text{RO-CH}_2\text{-R'} \xrightarrow{-2e,\ -H^+} \left[\text{RO-}\overset{+}{\text{C}}\text{H-R'} \longleftrightarrow \text{R}\overset{+}{\text{O}}\text{=CH-R'}\right] \xrightarrow{\text{Nu}^-} \text{RO-CH(Nu)-R'} \tag{4.102}$$

(Nu = MeO, AcO, F)

Reduction of Alcohol and Carboxylic Acid: One-electron reduction of these compounds generates the corresponding anions, eliminating a hydrogen molecule.

$$\mathrm{YOH} \xrightarrow{-e,\,-1/2H_2} \mathrm{YO^-}$$

$$(\mathrm{Y{=}Ar, R, RCO}) \qquad (4.103)$$

Reduction of Dioxygen: One-electron reduction of dioxygen generates superoxide ion, which is called active oxygen and works as both oxidant and reductant as well as base and radical (see section 5.2).

$$O_2 \xrightarrow{e} O_2^{\bullet -} \qquad (4.104)$$

4.7.2.3 Calcogeno (Sulfur, Selenium, Tellurium) Species

Carcogen compounds are easily oxidized and oxidation of the corresponding anion generates radicals that are further oxidized to cations, as shown in Eq. 4.105. These oxidation processes are usually reversible electron transfers although the reversibility depends on the electrolytic conditions.

$$\mathrm{RZ^-} \underset{e}{\overset{-e}{\rightleftarrows}} \mathrm{RZ^\bullet} \underset{e}{\overset{-e}{\rightleftarrows}} \mathrm{RZ^+} \qquad (4.105)$$

$$(\mathrm{Z{=}S, Se, Te})$$

In the case of sulfides, the anodically generated radical cation at the sulfur atom is attacked by nucleophiles (nucleophilic addition). However, in the case of sulfide with a strongly acidic α-hydrogen, α-deprotonation of the radical cation occurs predominantly and eventually α-nucleophilic substitution takes place selectively. Among the substitution reactions, fluorination is of great importance in synthetic organic chemistry. However, in the case of selenides, even selenides having an electron-withdrawing group do not always undergo nucleophilic substitutions, and a nucleophilic addition reaction also occurs at the selenium atom. In contrast, anodic substitution of organotelluides is not known and nucleophilic addition at the tellurium atom predominantly occurs.

$$\mathrm{RZCH_2{-}EWG} \xrightarrow{-2e,\,-H^+} \mathrm{R\overset{+}{Z}{=}CH{-}EWG} \xrightarrow[-H^+]{NuH} \mathrm{RZ\underset{Nu}{\underset{|}{C}}H{-}EWG} \tag{4.106}$$

(EWG: electron-withdrawing group)

(Z = S, Se)

(Nu = MeO, AcO, F)

4.7.2.4 Halogen Species

Halogen is easily oxidized and also readily reduced. Since the redox reaction of the halide ion is reversible, it is widely used as a mediator for indirect electrochemical oxidation (see Chapters 3 and 5).

$$X^- \underset{e}{\overset{-e}{\rightleftarrows}} X^\bullet \underset{e}{\overset{-e}{\rightleftarrows}} X^+ \tag{4.107}$$

(Z=Cl, Br, I)

4.7.2.5 14-Family and 15-Family Element Species

Reactive species are generated predominantly at heteroatoms by electron transfer, as shown in Eqs. 4.108–4.110.

$$R_3Y{-}X \xrightarrow{2e,\,-X^-} R_3Y^- \tag{4.108}$$

(Y=Si, Sn)

$$R_2Z{-}X \xrightarrow{2e,\,-X^-} R_2Z^- \tag{4.109}$$

$$R_3Z \xrightarrow{-e} R_3Z^{\cdot+} \tag{4.110}$$

(Z=P, As, Sb)

REFERENCES

1. Torii, S. (2006) *Electroorganic Reduction Synthesis*, Vols 1 and 2, Kodansha and Wiley-VCH Verlag GmbH, Weinheim.
2. Bard, A.J. and Stratmann, M. (eds) (2002) Organic electrochemistry, in *Encyclopedia of Electrochemistry*, Vol. 8 (ed. H. J. Schäfer), *Organic Electrochemistry* Wiley-VCH Verlag GmbH, Weinheim.

3. Lund, H,. and Hammerich, O. (eds) (2001) *Organic Electrochemistry*, 4th edn, Marcel Dekker, Inc., New York.
4. Grimshaw, J. (2000) *Electrochemical Reactions and Mechansims in Organic Chemistry*, Elsevier, Amsterdam.
5. Fry, A. J. (1989) *Synthetic Organic Electrochemistry*, Wiley Interscience, New York.
6. Shono, T. (1984) *Electroorganic Chemistry as a Tool in Organic Synthesis*, Springer-Verlag, Berlin.
7. Fuchigami, T., Nonaka, T. and Schäfer, H.J. (2003) *Encyclopedia of Electrochemistry*, Vol. 8 (eds A. J. Bard and M. Stratmann), Wiley-VCH, Verlag GmbH, Weinheim.
8. Fussing, I., Hammerich, O., Hussain, A., Nielsen, M.F. and Utley, J.H.P. (1998) *Acta Chem. Scand.*, **52**, 328–171.
9. Fuchigami, T., Ichikawa, S. and Konno, A. (1994) *J. Org. Chem.*, **59**, 607–615.
10. (a) Nelsen, S.F. and Ippoliti, J.T. *J. Am. Chem. Soc.*, **108**, 4879–4881. (b) Yoon, V.C. and Mariano, P.S. (1992) *Acc. Chem. Res.*, **25**, 233–240.
11. Fuchigami, T., Narizuka, S. and Konno, A. (1992) *J. Org. Chem.*, **57**, 3755–3757.
12. Higashiya, S., Narizuka, S., Konno, A. and Fuchigami, T. (1992) *J. Org. Chem.*, **57**, 3755–3757.
13. Edamura, F., Kyriyacou, D., Love, J. (1980) US Patent 4217185; *Chem. Abstr.* (1981), 94, 22193.
14. (a) Tokuda, M., Yamada, Y., Takagi, T., Suginome, H. and Furusaki, A. (1985) *Tetrahedron Lett.*, **26**, 6086–6089. (b) Tokuda, M., Yamada, Y., Takagi, T., Suginome, H. and Furusaki, A. (1987) *Tetrahedron*, **43**, 281–296.
15. (a) Fry, A.J. and Moor, R.H. *J. Org. Chem.*, **33**, 1283–1284. (b) Erickson, R.E., Annino, R., Sainlor, M.D. and Zon, G. (1969) *J. Am. Chem. Soc.*, **91**, 1767–1770.
16. Sekine, T., Yamura, A. and Sugino, K. (1965) *J. Electrochem. Soc.*, **112**, 439–443.
17. Nonaka, T., Wachi, S. and Fuchigami, T. (1977) *Chem. Lett.*, 47–50.
18. (a) Yoshida, J., Kataoka, K., Horcajada, R. and Nagaki, A. (2008) *Chem. Rev.*, **108**, 2265–2299. (b) Yoshida, J. and Nishiwaki, K. (1998) *J. Chem. Soc., Dalton Trans.*, 2589–2596. (c) Yoshida, J., Maekawa, T., Murata, T., Matsunaga, S. and Isoe, S. (1990) *J. Am. Chem. Soc.*, **112**, 1962–1970.
19. Koizumi, T., Fuchigami, T. and Nonaka, T. (1989) *Bul. Chem. Soc. Jpn*, **62**, 219–225.
20. Yoshida, J. (1994) Electrochemical Reactions of Organosilicon Compounds, in *Topics in Current Chemistry, 170. Electrochemistry V*. Springer-Verlag, Berlin, pp. 39–82.
21. Fuchigami, T. (1998) Electrochemistry of Organosilicon Compounds, in *The Chemistry of Organic Silicon Compounds*, Vol. 2 (eds Z. Rappoport and Y. Apeloig), John Wiley & Sons, Ltd, Chichester, Chapter 20.
22. Gall, E.L., Hurvois, J.P. and Sinbandhit, S. (1999) *Eur. J. Org. Chem.*, 2645–2653.
23. (a) Kim, S., Hayashi, K., Kitano, Y. and Chiba, K. (2002) *Org. Lett.*, **4**, 3735–3737. (b) Chiba, K., Uchiyama, T., Kim, S., Kitano, Y. and Tada, M. (2001) *Org. Lett.*, **3**, 1245–1248.
24. Kandeel, Z., Nonaka, T. and Fuchigami, T. (1986) *Bull. Chem. Soc. Jpn*, **59**, 338–340.
25. (a) Uneyama, K. and Kato, T. (1998) *Tetrahedron Lett.*, **39**, 587–590. (b) Uneyama, K., Naeda, K., Kato, T. and Katagiri, T. (1998) *Tetrahedron Lett.*, **39**, 3741–3744.
26. (a) Yoshida, J. and Izawa, M. (1997) *J. Am. Chem. Soc.*, **119**, 9361–9365. (b) Watanabe, M., Suga, S. and Yoshida, J. (2000) *Bull. Chem. Soc. Jpn*, **73**, 243–247. (c) Yoshida, J., Suga, S., Fuke, K. and Watanabe, M. (1999) *Chem. Lett.*, 251–252.

27. Schäfer, H. (1990) Recent Contributions of Kolbe Electrolysis to Organic Synthesis, in *Topics in Current Chemistry, 152. Electrochemistry, IV* (ed. E. Steckhan), Springer-Verlag, Berlin, pp. 91–151.
28. (a) Fuchigami, T. and Nonaka, T. (1984) *Electrochemistry*, **52**, 19–25. (b) Fuchigami, T., Sato, T. and Nonaka, T. (1986) *J. Org. Chem.*, **51**, 366–369.
29. Baba, D. and Fuchigami, T. (2003) *Tetrahedron Lett.*, **44**, 3133–3136.
30. Fuchigami, T., Iwata, K. and Nonaka, T. (1976) *J. Chem. Soc., Chem. Commun.*, 951–952.
31. Gassman, P.G., Nishiguchi, I. and Yamamoto, H. (1975) *J. Am. Chem. Soc.*, **97**, 1600–1602.
32. Fuchigami, T., Sato, T. and Nonaka, T. (1986) *Electrochim. Acta*, **31**, 365–369.
33. Watson, L.D., Yu, L. and Yudin, A.K. (2006) *Acc. Chem. Res.*, **39**, 194–206.

5

Organic Electrosynthesis

Toshio Fuchigami and Shinsuke Inagi

Development of new methodology to achieve highly selective reactions is one of the most important areas in organic synthetic chemistry. As already described, electrode reactions have their own specific factors for controlling selectivity, therefore both electrochemical and ordinary chemical factors make the control of electrochemical reactions more complicated. Electrodes are of great importance for both electron transfer interfaces and reaction fields. As described earlier, an electrode has a function to control a chemical reaction pathway through adsorption and orientation of the substrate molecule to the electrode surface. Although hydrogen and oxygen overpotentials could be criteria for the selection of suitable electrode materials to achieve the desired electrochemical reaction in an aqueous solution, these overpotentials are not proper criteria in aprotic solutions. Hence, it is not so easy to predict suitable electrode materials for desired electrochemical reaction in aprotic solvents. However, many novel electrolytic methodologies have been developed in order to achieve high selectivity for the desired reactions. In this chapter, relatively new electrolytic methodologies, which have already been established and are widely used, will be described in detail. Although there are many applications of such electrochemical reactions to organic syntheses, limited

Fundamentals and Applications of Organic Electrochemistry: Synthesis, Materials, Devices, First Edition. Toshio Fuchigami, Mahito Atobe and Shinsuke Inagi.

examples are given in this chapter. It is recommended that readers also study other more detailed books [1].

5.1 ELECTROCATALYSIS

Although indirect electrochemical reactions using mediators, i.e. electrocatalytic reactions, have already been explained in Chapter 2, various examples of synthetic applications are described in this chapter [2]. Some other synthetic examples are also demonstrated in Chapters 6 and 7.

5.1.1 Classification and Kinds of Mediators

Mediators are classified into two groups, as shown in Eqs. 5.1 and 5.2: outer sphere type mediators involving electron transfer between the mediator and a substrate molecule (Eq. 5.1), and inner sphere type mediators involving redox reaction through ordinary chemical reactions (Eq. 5.2), as shown in Table 5.1.

R–X ⟶ R• + X^-

e

(5.1)

I^-

–2e

I^+

R^1R^2CHOH ⟶ (KI) R^1R^2CHOI ⟶ (–HI, KOH) $R^1R^2C{=}O$

(5.2)

Table 5.1 Classification of mediators

Mediator	
Outer sphere	Inner sphere
Multivalent metal ion	Halogen
Triarylamine	*N*-Oxyl compound
Viologen	Sulfide
Polycyclic aromatic hydrocarbon	

Mediators are also classified into oxidative and reductive mediators, and into organic and inorganic mediators. Various redox mediators are known, for example (i) inorganic compounds such as multivalent metal ions, transition metal complexes and halide ions, (ii) organic compounds such as polycyclic aromatic compounds, triarylamines and 2,2,6,6-tetramethylpiperidine nitroxyl (TEMPO), (iii) anomalous valence compounds and (iv) hypervalent compounds. Quite recently, novel mediators such as carboranes, which have the characteristics of boron atoms, have been developed for highly efficient cathodic reductive dehalogenation [3]. Triarylimidazole mediators have also recently been developed for selective anodic oxidation [4].

Examples of oxidative and reductive mediators are as follows:

1. **Oxidative Mediators**
   Ru^{4+}/Ru^{2+}, Co^{3+}/Co^{2+}, IO_4^-/IO_3^-, Mn^{3+}/Mn^{2+}, Ni^{3+}/Ni^{2+}, Ce^{4+}/Ce^{3+}, $S_2O_8^{2+}/SO_4^{2+}$, Cr^{6+}/Cr^{3+}, Os^{8+}/Os^{6+}, halogen (X^+/X^-), $NO_3^{\cdot}/NO_3^-$, CAN, Ar_3N^+/Ar_3N, *N*-Oxyl derivatives ($R^1R^2N{=}O$), sulfides (R^1SR^2).
2. **Reductive Mediators**
   $Ni^{2+}/Ni(0)$, Co^{2+}/Co^+, Cr^{3+}/Cr^{2+}, $Mo^{2+}/Mo(0)$, Ti^{3+}/Ti^{2+}, Sn^{4+}/Sn^{2+}, $Sn^{2+}/Sn(0)$, $Mg^{2+}/Mg(0)$, $Pd^{2+}/Pd(0)$, naphthalene, anthracene, phenathrene, pyrene, fullerene C_{60}, benzonitrile, phthalonitrile, antraquinone, nitrobenzene, viologen, superoxide ion.

The reduction potentials of aromatic mediators for indirect cathodic reduction are shown in Table 5.2.

Table 5.2 Reduction potentials of aromatic mediators for indirect cathodic reduction

Compound	$E_{1/2}$ (V vs. SCE)
Phthalonitrile	−1.7
4-Methoxybenzophenone	−1.8
Anthracene	−2.0
Methyl benzoate	−2.2
Benzonitrile	−2.2
Chrysene	−2.5
Naphthalene	−2.5

5.1.2 Organic Electrolytic Reactions Using Mediators

5.1.2.1 Electrosynthesis Using Multivalent Metal Ion Mediators

Polyvalent metal ions have been used for a long time and some are still used for industrial electrolysis. Synthetic examples using such mediators are illustrated in Eq. 5.3. Direct electrolysis of toluene derivatives provides a mixture of aldehyde and carboxylic acid derivatives, while each product can be obtained selectively by the choice of appropriate mediator.

R–C₆H₄–CH₃ → (Direct anodic oxidation) → R–C₆H₄–CHO + R–C₆H₄–COOH

R–C₆H₄–CH₃ → Ce^{4+}/Ce^{3+}, Mn^{3+}/Mn^{2+}, or $S_2O_8^{2+}/SO_4^{2+}$ (Indirect anodic oxidation) → R–C₆H₄–CHO

R–C₆H₄–CH₃ → Cr^{6+}/Cr^{3+} (Indirect anodic oxidation) → R–C₆H₄–COOH

(5.3)

5.1.2.2 Electrosynthesis Using Halogen Mediators

Anodic oxidation of halide ions (X^-) generates various kinds of reactive cationic species, for example X^+, OX^-, X_2, X_3^- etc., which are widely utilized as mediators for various oxidative molecular conversions, as shown in Figure 5.1. Oxidative power increases in the order I < Br < Cl,

R^1R^2CH–OH → (KI/H_2O/*t*-BuOH; I^+ / I^-; –2e Anode) → R^1R^2C=O

$R^1 = R^2 =$ (bicyclic): 93%

$R^1 = C_6H_{13}$, R^2 = Me: 99%

$R^1 = R^2$ = Cyclohexyl: 74%

R^1 = Ph, R^2 = Me: 100%

Figure 5.1 Indirect anodic oxidation of alcohols using iodine mediator

and positive Cl is too strong oxidant to give good selectivity. Iodide and bromide ions are therefore mainly employed as mediators for selective indirect oxidation. The synthetic applications of such electrogenerated active halogen species as mediators have been demonstrated by highly selective functionalization of olefins, e.g. epoxidation, halohydroxylation, 1,2-dihalogenation and ene-type chlorination, heteroatom–heteroatom bond formation and carbon–heteroatom bond cleavage [5–8].

5.1.2.3 Electrosynthesis Using Triarylamine Mediators

Triarylamines, which are well known as outer sphere electron transfer reagents, were first used as mediators for anodic oxidative deprotection of a protective group like dithiolane for a carbonyl group by Steckhan (Eq. 5.4) [2]. The oxidation ability of this mediator can be tuned by substitution with electron-withdrawing groups such as the bromine atom, as shown in Table 5.3. Triarylamine mediators can be applied to oxidation of alcohols, amines and the side-chain of aromatic compounds, together with fluorodesulfurization. Thus, they are highly useful and widely applicable mediators.

$$\text{RCH(dithiane)} \xrightarrow[Na_2CO_3\ /aq.MeCN]{2A^{\bullet +} \rightleftarrows 2A,\ 2e} RCHO + \text{(1,2-dithiolane, S–S)} \quad (5.4)$$

Table 5.3 Standard oxidation potentials of triarylamines

Triarylamine	Substituent			E^o (V vs. NHE)
	X	Y	Z	
A	Br	H	H	1.30
B	Br	Br	H	1.74
C	Br	Br	Br	1.96

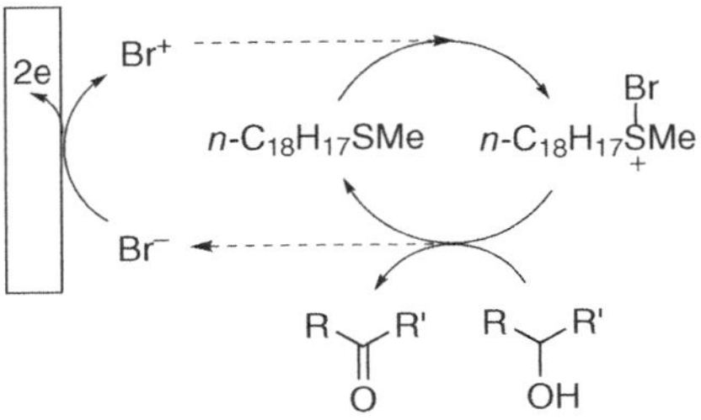

Figure 5.2 Indirect anodic oxidation of alcohols using double mediator consisting of sulfide and bromide ion

5.1.2.4 Electrosynthesis Using Multi-Mediatory Systems

Various combinations of different kinds of mediators have been developed to expand the scope of electrosynthesis. As shown in Figure 5.2, a double mediator consisted of sulfide and bromide ions enables the oxidation of alcohols at much lower oxidation potential [9].

5.1.2.5 Electrosynthesis Using Hypervalent Compounds as Mediators

Anodic oxidation of *p*-methoxyiodobenzene in the presence of fluoride ions provides the corresponding hypervalent difluoroiodo derivative, which is a useful fluorinating reagent. Thus, *p*-methoxyiodobenzene has been demonstrated to be a highly efficient mediator for anodic fluorodesulfurization of dithioacetals as shown in Figure 5.3 [10]. This is the first successful example of the catalytic use of hypervalent compounds for organic synthesis as well as organic electrosynthesis.

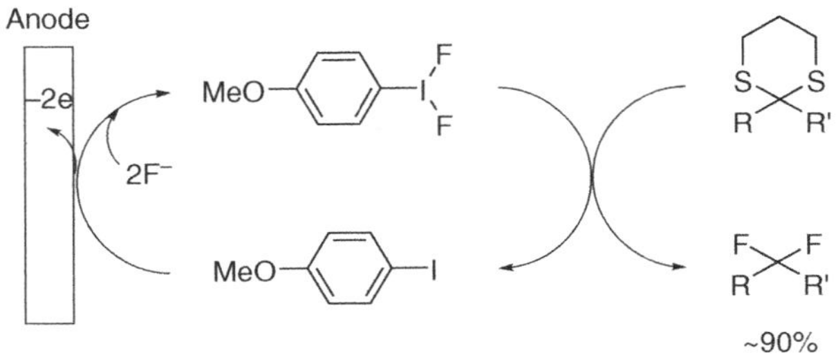

Figure 5.3 Electrochemical fluorodesulfurization using hypervalent iodobenzene derivative mediator

Anodic α-fluorination of α,γ-diketones and α-ketoesters can be also achieved using *p*-iodotoluene as a mediator [11].

5.1.2.6 Electrosynthesis Using Transition Metal Complex Mediators

Transition metal complex mediators have various reactivities and many advantages. Their redox potentials and the selectivity of the desired reaction can be controlled by changing ligand. Their synthetic application is mainly based on the reactivity of the low valent state generated by cathodic reduction of the mediators [2]. Typically, Co(III) complex (vitamin B_{12}) is readily reducible, optically active, non-toxic and inexpensive. It is reduced at −0.9 V vs. SCE to form Co(I) complex, which undergoes oxidative addition to alkyl halide to form alkyl Co(III) complex as an intermediate. The resulting intermediate is reduced at more negative potential, −1.5 V vs. SCE, to generate an alkyl radical or anion, and Co(I) complex is regenerated simultaneously (reductive elimination). Since the resulting alkyl radical or anion undergoes conjugate addition, efficient Michael addition can be achieved under neutral conditions, as shown in Figure 5.4 [12]. This mediatory reaction is widely applicable to various halogeno compounds, such as allyl halides, vinyl halides, α-halo ethers and so on.

As shown in Eq. 5.5, even halogeno compounds with non-protected hydroxyl and carbonyl groups can be used, which is one of the advantages of this mediatory system [13].

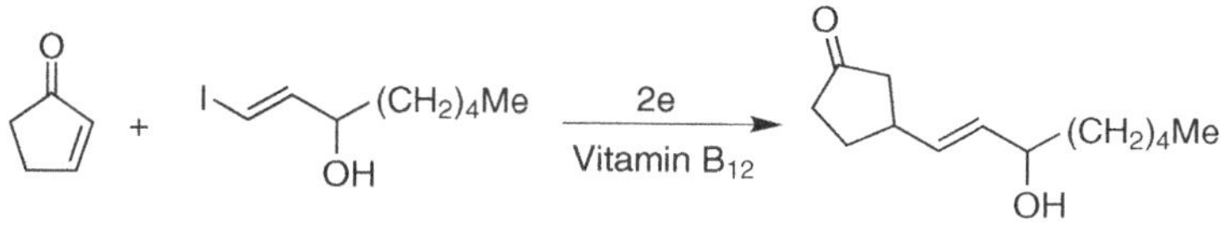

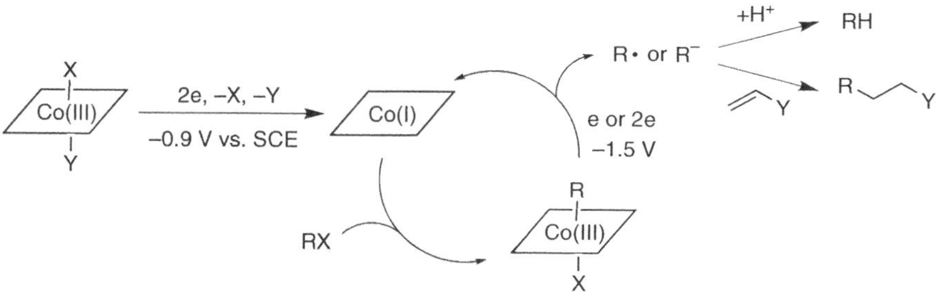

Figure 5.4 Electrosynthesis using Co(III) complex, vitamin B_{12} as mediator

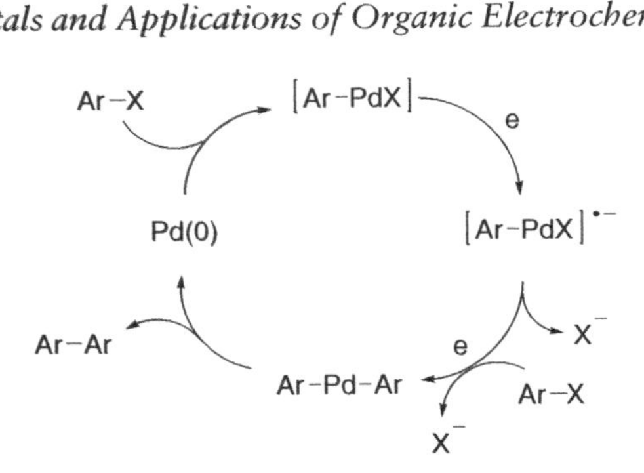

Figure 5.5 Homo coupling of aryl halide using Pd(0) complex

On the other hand, β-bromodiester and ω-bromoalkyl acrylate undergo 1,2-rearrangement and intramolecular cyclization, respectively, to form large cyclic lactones by their cathodic reduction using hydrophobic vitamin B_{12} mediator under UV irradiation [14].

Homo-coupling products are obtained from aromatic halides using Pd(0) complex as well as Ni(0) complex as a mediator [15]. The yields and turnover of the Pd(0) complex are generally superior to those using Ni(0) complex. The mechanism proposed is shown in Figure 5.5. Aromatic halide reacts with Pd(0) complex to generate an aryl Pd intermediate, which is reduced cathodically followed by reaction with one more aryl halide molecule to form diaryl Pd complex, resulting in reductive elimination to give a homo-coupling product. When this reaction is performed in the presence of CO_2, aromatic carboxylic acids are obtained in high yields [16].

5.1.2.7 Electrosynthesis Using Mediator Immobilzed on Solid

N-Oxyl radicals exemplified by 2,2,3,3-tetramethylpiperidinyl-*N*-oxyl (TEMPO) are useful mediators for efficient oxidation of alcohols and amines. The mediators have never been recovered and reused after electrolysis, but recyclable mediators have recently been developed from an atom economical aspect. Mediators are immobilized on the surface of silica gel through silane coupler, and are readily recovered and recycled after electrolysis. Thus, the combination of halide ions and TEMPO-immobilized silica gel or polymer particles as a disperse phase enables a mediated electrocatalytic reaction in aqueous NaBr-$NaHCO_3$ as a disperse media [17]. Asymmetric oxidation of alcohols is also possible using a chiral *N*-oxyl mediator [18].

5.2 ELECTROGENERATED ACIDS AND BASES

5.2.1 Electrogenerated Bases

Acids and bases play very important roles in organic synthesis. It is well known that when aqueous solution containing a neutral supporting electrolyte is electrolysed in a divided cell, the anolyte becomes acidic while the catholyte becomes alkaline. This is because hydroxide ions are consumed at the anode while protons are consumed at the cathode. In a similar manner, some acid is generated in the anolyte while some base is generated in the catholyte during electrolysis in an organic solvent. Even in an undivided cell, the vicinity of an anode becomes acidic while that of a cathode becomes basic during electrolysis.

Anionic species generated cathodically act not only as nucleophiles but also as bases, and have interesting reactivities in organic synthesis. The inventor of the cathodic hydrodimerization process of acrylonitrile, Baizer, demonstrated that the cathodically generated anion radical of hindered azobenzene (Figure 5.6) could be a useful base for various organic synthesis, and he named such bases electrogenerated bases (EGBs) [19].

There are two main methods for the generation of EGBs, as shown in Eq. 5.6. One method is cathodic reduction of compounds with an unsaturated moiety in aprotic solvents containing quaternary ammonium salt as a supporting electrolyte to generate the radical anions or dianions, and the other one is cathodic reductive deprotonation of active hydrogen compounds to generate the corresponding anions.

$$\begin{gathered} Y=Z \xrightarrow[R_4N^+X^-]{e} [Y=Z]^{\bullet-}R_4N^+ \\ Y-H \xrightarrow[R_4N^+X^-]{e,\,-1/2\,H_2} Y^-R_4N^+ \\ (Y, Z = C, O, S, N, P) \end{gathered} \tag{5.6}$$

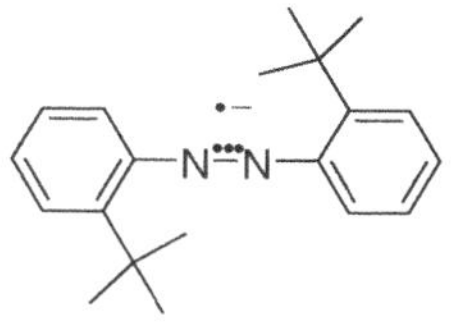

Figure 5.6 Electrogenerated base of hindered azobenzene

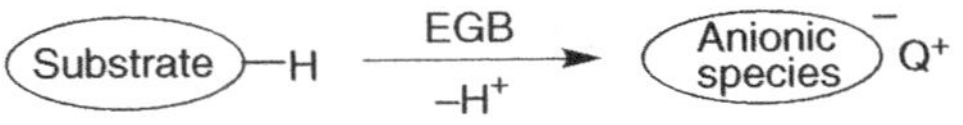

Substrate—H —EGB, $-H^+$→ Anionic species$^-$ Q^+

Figure 5.7 Reactive anion derived from electrogenerated base ($Q^+ = Et_4N^+$)

Cathodic reductive deprotonation of 2-pyrrolidone, a hindered phenol like 2,6-*t*-butyl-4-methylphenol, and triphenylmethane generates the corresponding anions. Since EGBs have quaternary ammoniumcations (Q^+), the anions formed by the treatment with EGB have high reactivity (Figure 5.7) and hence this methodology is highly useful for organic synthesis.

In particular, the EGB derived from 2-pyrrolidone is a versatile base and applicable to organic synthesis such as Stevens rearrangement, selective α-monoalkylation of α-(aryl)acetate esters and C-monoalkylation of 1,3-diketones [20,21]. This EGB has also been demonstrated to be a highly efficient base for the synthesis of organofluorine compounds. For instance, it is known that trifluoromethyl anion is so unstable that it undergoes α-elimination of fluoride anion to generate fluorocarbene, but stable trifluoromethyl anion can be generated by the treatment of fluoroform with this EGB, and consequently trifluoromethylation of aromatic aldehydes and ketones is realized to give the trifluoromethylated alcohols, as shown in Eq. 5.7 [22].

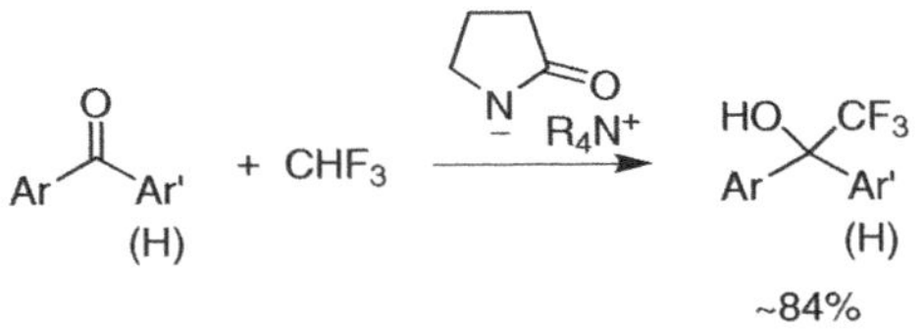

$$ArC(=O)Ar'(H) + CHF_3 \xrightarrow{\text{pyrrolidone anion } R_4N^+} ArC(OH)(CF_3)Ar'(H) \quad \sim 84\% \tag{5.7}$$

Generally, it is quite difficult to generate the enolate anion with an α-CF_3 group since the enolate anion readily undergoes decomposition such as β-elimination of fluoride ion. However, pyrrolidone-derived EGB enables the generation of stable trifluoromethylated enolate anion, and alkylation can be performed in good yield without elimination of the fluorine atom, as shown in Eq. 5.8 [23].

e
$-1/2\ H_2$
Et_4NOTs
Et_4N^+
DMF
0°C
MeO OMe
CF_3
Et_4N^+
RX
~ r.t.
R CF_3

R = Me : 80%
R = $PhCH_2$: 63%
R = $CH_2{=}CHCH_2$: 70%

(5.8)

Furthermore, this EGB catalyzes ring-opening polymerization of N-carboxyanhydrides of α-amino acid (NCA) to provide poly(amino acids) in a short time in excellent yield, as shown in Eq. 5.9. The yield is much higher and the reaction time relatively short compared to reactions using conventional base with a metal cation such as Na^+ [24].

L-valine NCA
Q^+
$-H^+$
NCA
$-CO_2$
$-(NHCHCO)_n-$

$Q^+ = Bu_4N^+$: 98%
$Q^+ = Na^+$: 7%

(5.9)

EGB derived from hindered phenol catalyzes a selective double aldol condensation reaction, as shown in Eq. 5.10 [25].

2 Ar CH_3 + Ar' H
Et_4N^+
Ar Ar' Ar
~ 90%

(5.10)

It is well known that the substrate molecule itself also serves as an EGB. For instance, as shown in Eq. 5.11, a catalytic amount of EGB efficiently catalyzes aldol condensation. In this reaction, aldehyde itself is cathodically reduced to generate a trace amount of its radical anion, which acts as a base to abstract the α-proton from unreacted aldehyde, and aldol condensation proceeds successively [26]. Finally, eliminating hydroxyl ion would act as an EGB in a manner similar to a chain reaction.

$$\text{CH}_3\text{CH}_2\text{CH}_2\text{CHO} \xrightarrow[\text{DMF}]{e} [\text{CH}_3\text{CH}_2\text{CH}_2\text{CHO}]^{\bullet -} \equiv \text{EGB}$$

$$\text{CH}_3\text{CH}_2\text{CH}_2\text{CHO} \xrightarrow[-\text{H}^+]{\text{EGB}} \text{CH}_3\text{CH}_2\text{CH}^{-}\text{CHO}$$

$$\xrightarrow{\text{CH}_3\text{CH}_2\text{CH}_2\text{CHO}} \text{CH}_3\text{CH}_2\text{CH}_2\text{CH}(\text{O}^-)\text{CH}(\text{CH}_2\text{CH}_3)\text{CHO} \xrightarrow{-\text{OH}^-\ (\text{EGB})} \text{CH}_3\text{CH}_2\text{CH}_2\text{CH}=\text{C}(\text{CH}_2\text{CH}_3)\text{CHO}$$

Yield : 76%
Current efficiency : 9 x 10^3%

(5.11)

Treatment of alcohols with sodium metal generates hydrogen gas and sodium alkoxide is formed. This can be explained as follows. The peripheral electron of sodium metal transfers to the alcohol to generate a hydrogen radical, forming hydrogen gas, and the alcohol is transformed to the alkoxide, in other words sodium metal acts as a reducing reagent for the alcohol. When the alcohol is reduced at a cathode with a low hydrogen overpotential, like platinum, hydrogen gas evolves and the corresponding alkoxide is readily formed without any reducing reagent. It can therefore be considered that the cathode would act as sodium metal. This is quite a safe method for the preparation of alkoxide because harmful sodium metal is not required. Moreover, the cathodic reduction of carboxylic acid, phenol, thiophenol and alcohol is performed in the presence of quaternary ammonium salt as a supporting electrolyte to generate the corresponding highly reactive anions with quaternary ammonium cation (Q^+). By using such reactive anions, highly selective esterification, etherification and carbonylation can be achieved at room temperature, as shown in Eq. 5.12 [27,28].

$$\text{RCOOH} \xrightarrow{e\ -1/2\text{H}_2} \text{RCOO}^-\text{Q}^+ \xrightarrow{\text{R'X}} \text{RCOOR'}\ (\sim 96\%)$$

$$\text{ArYH} \xrightarrow{e\ -1/2\text{H}_2} \text{ArY}^-\text{Q}^+ \xrightarrow[(\text{Y = O, S})]{\text{Ar'X}} \text{ArYAr'}\ (\sim 98\%) \qquad (\text{Q}^+=\text{Et}_4\text{N}^+) \qquad (5.12)$$

$$\text{ROH} \xrightarrow{e\ -1/2\text{H}_2} \text{RO}^-\text{Q}^+ \xrightarrow{\text{Fe(CO)}_5} \text{ROCO}^- \xrightarrow{\text{R'X}} \text{ROCOR'}$$

Cathodic reduction of dioxygen generates superoxide ion ($O_2^{-\bullet}$). Superoxide ion acts as nucleophile, oxidant, reductant, radical and base (EGB), therefore this reactive species is highly useful for organic synthesis. Although superoxide ions are available from KO_2 and NaO_2, these are insoluble in aprotic solvents. In order to dissolve these salts in aprotic solvents, costly crown ethers are required therefore their synthetic applications are limited. On the other hand, the electrolytic method can generate superoxide ion *in situ*, which is better than the conventional chemical method. As shown in Eq. 5.13, electrogenerated superoxide ion acts as an EGB to eliminate the α-proton of malonic ester derivative, followed by reaction with O_2 to give the α-hydroxy product in good yield [29]. High current efficiency suggests that this reaction may involve a chain reaction.

$$CH_3CH(COOEt)_2 \xrightarrow{O_2^{\bullet-}} CH_3C(OH)(COOEt)_2 \quad (95\%) \tag{5.13}$$

5.2.2 Electrogenerated Acids

In contrast to EGBs, the vicinity of an anode becomes acidic during electrolysis. The acid thus formed is called an electrogenerated acid (EGA), which has unique reactivity compared to conventional chemical acids. For instance, electrolysis of a solution containing $LiClO_4$ and a trace amount of water generates acid, $HClO_4$, as shown in Figure 5.8 [30].

EGA generated in this way is assumed to be anhydrous $HClO_4$, which is a stronger acid than commercially available aq. $HClO_4$. Such EGA acts as a kind of Lewis acid and is widely applicable to various organic reactions,

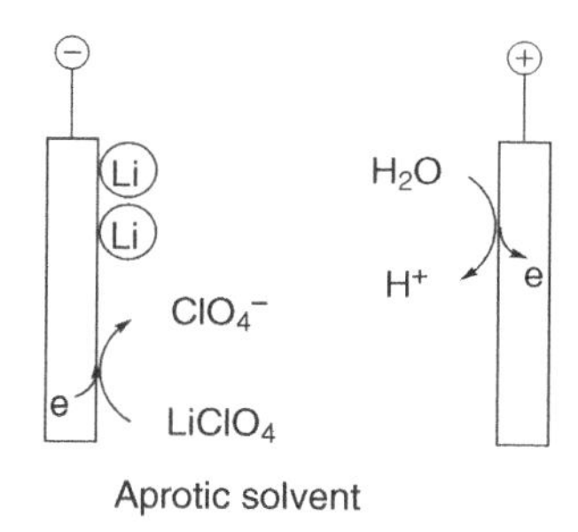

Figure 5.8 Principle of electrogenerated acid

for instance isomerization, transformation of functional groups, carbon–carbon bond formation, including Diels–Alder reaction, and so on [31]. As shown in Eq. 5.14, a combination of supporting salts and solvents allows three kinds of products to be obtained selectively from the same starting material [30].

EGA
$LiClO_4/CH_2Cl_2$

EGA
$Mg(ClO_4)_2/(CH_3)_2CO$

EGA
$Et_4NOTs+NaOTs$
$/ClCH_2CH_2Cl$

R O R O O R OH

(5.14)

Furthermore, EGAs are also highly effective for the molecular transformation of organofluorine compounds. For instance, as shown in Eq. 5.15, the α-cation attached to the CF_3 group is catalytically generated by the treatment of trifluoromethylated *O,S*-acetal with an EGA, and subsequently carbon nucleophiles are readily introduced to the α-position [32]. However, use of other conventional Lewis acids results in no formation of the desired product.

$$PhSCH(OAc)CH_2CF_3 \xrightarrow[Nu\text{-}SiMe_3]{EGA} PhSCH(Nu)CH_2CF_3 \qquad (5.15)$$

Nu = CN : 56%
Nu = $CH_2CH{=}CH_2$: 88%

5.3 ELECTROCHEMICAL ASYMMETRIC SYNTHESIS

Asymmetric synthesis is a highly important area in organic chemistry and a number of asymmetric catalytic syntheses have recently been developed. However, electrochemical asymmetric synthesis is still immature compared with well-established chemical, catalytic and enzymatic methods. Various methods for electrochemical asymmetric synthesis have been proposed [33,34]: (i) intramolecular asymmetric induction, (ii) utilization of chiral solvents, (iii) use of chiral supporting salts, (iv) use of chiral electrode adsorbance, (v) use of electrodes chemically modified with chiral substances, (vi) use of chiral polymer-modified electrodes, (vii) use of chiral mediators. Among these, (i) and (ii) are not unique methodology characterized by electrochemistry. However, a number of papers dealing with these methodologies have been reported. Although (iii) is a methodology characterized by electrolysis, a large amount of chiral supporting salt is required, therefore it is not useful. Here, methods (iv)–(vi) are explained.

In 1967, Grimshaw first reported the asymmetric reduction of 4-methylcumarine using optically active alkaloids, and a maximum asymmetric yield (17%) was obtained using a small amount of sparteine (1 mM) as a chiral electrode adsorbant [35]. After 10 years, Miller obtained 48% asymmetric yield in the cathodic reduction of 2-acethylpyridine to the corresponding alcohol in the presence of 0.5 mM of strychnine salt [36]. Asymmetric induction seems to be attributable to the chiral reaction fields constructed by the physical adsorption of optically active alkaloids on the cathode surface. However, such asymmetric electrosynthesis is very sensitive to electrolytic conditions such as stirring, concentration, solvent, cathode potential and so on.

Miller and co-workers found that prochiral carbonyl compounds like ethyl phenylglyoxylate were reduced to chiral alcohols in 10% asymmetric yield on a cathode chemically modified with (*S*)-phenylglycine [37]. They also reported asymmetric oxidation of *p*-tolyl methyl sulfide to the sulfone with 2.5% ee by using a similar electrode modified with (+)-camphoric acid [38]. These are also pioneering works on chemically modified electrodes. The low asymmetric yields seem to be due to a low density of chiral compound on the electrode surface, and other researchers have pointed out that Miller's results are not always reproducible. Later on, in order to solve problems such as the instability of electrodes modified with chiral adsorbance and the low density of modified chiral compounds, chiral polymer-modified electrodes were developed. For instance, Nonaka, Fuchigami and co-workers prepared various electrodes coated with optically active poly(amino acid)s and applied these to the asymmetric reduction of olefins. They obtained 43% ee in the reduction of 4-methylcumarin at a cathode coated with poly(*L*-valine) as shown in Figure 5.9 [39].

Subsequently, a graphite felt electrode modified with 2,2,6,6-tetramethylpiperidin-1-yloxyl (TEMPO) was developed and applied to enantioselective oxidative coupling of 2-naphthol, 2-methoxynaphthalene and 10-hydroxyphenanthrene in the presence of (−)-sparteine as a chiral

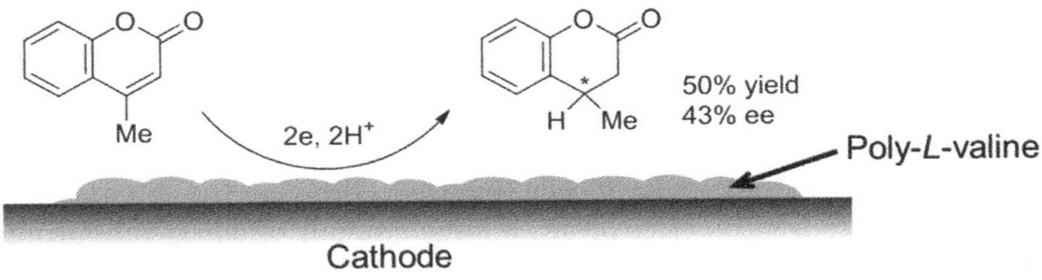

Figure 5.9 Asymmetric reduction using chiral polymer-modified electrode

base. The enantioselectivity of the coupling products was very high (98%) [40,41]. It has been considered that high asymmetric yield cannot be expected since electrolytic reactions usually require polar solvents. It should therefore be noted that such excellent results of asymmetric electrosynthesis were realized even in polar solvents.

The combination of enzymes such as lactic acid dehydrogenase and mediators or chiral mediators is also effective for electrochemical asymmetric synthesis and kinetic resolution to provide products with high asymmetric yields [42–44].

5.4 MODIFIED ELECTRODES

As described previously, electrode surfaces and the vicinities of electrodes are the main reaction fields for electrode reactions, therefore if the electrode surface is modified with desired functional substances such as electrocatalyst, and chiral compound, the efficiency, desired selectivity, asymmetric synthesis, and etc. would be realized. Such electrodes are called modified electrodes and their use enables the amount of functional materials required to be reduced and the work-up for electrolytic solutions to be simplified [45]. Excellent durability is needed for the modified electrodes for electrosynthesis since both a large electrolytic current and long electrolysis time are required, which is different from modified electrodes for sensors. Modified electrodes are classified as follows depending on modification methods and modifiers.

5.4.1 Electrodes Modified with Adsorbants

When the electrode is immersed in a solution containing redox compounds like flavins and phenazines, and then pulled up from the solution, an electrode modified with the redox compounds is readily obtained. The modification is not very strong, the durability of the modified electrode is not great and the modification density is low. On the other hand, an electrode modified with strongly adhering and homogeneous films with redox catalysts using the Langmuir–Blodgett technique is superior to an adsorbant-modified electrode since both the modification density (density of redox catalyst) and durability of the Langmuir–Blodgett film modified electrode are much higher. Based on hydrophilic/hydrophobic control of the electrode surface using Langmuir–Blodgett film containing quaternary ammonium salt, product selectivity control has been achieved [46]. Langmuir–Blodgett film-coated electrodes also allow electrocatalytic reactions to be achieved [47].

5.4.2 Foreign Metal Adatom Modified Electrodes

Base electrodes can be modified by deposition of foreign metal adatoms at potentials several hundred millivolts positive to the reversible potential for metal deposition. A submonolayer of adatoms covers the base electrode surface uniformly, making it into an adatoms modified electrode [48]. Such electrodes are highly useful for electrocatalysis of oxygen reduction and oxidation of methanol, therefore the electrodes are currently of great importance in the development of fuel cells.

These electrodes also allow interesting product selectivity as follows. Cathodic reduction of dinitrobenzenes at a silver cathode in an acidic methanol solution mainly provides diaminobenzenes, while the use of a silver cathode modified by deposition of lead adatoms forms phenylene dihydroxyamines selectively, as shown in Eq. 5.16 [49]. This is considered to be due to the inhibition of adsorption of the intermediate phenylene dihydroxyamines on the electrode surface by deposited lead atoms, which interfere with further reduction.

$$C_6H_4(NO_2)_2 \xrightarrow{8e,\ 8H^+} \underset{\text{Pb-Ag cathode}}{C_6H_4(NHOH)_2} \xrightarrow{4e,\ 4H^+} \underset{\text{Ag cathode}}{C_6H_4(NH_2)_2} \qquad (5.16)$$

5.4.3 Chemically Modified Electrodes

As described above, Miller's work on chiral compound modified electrodes has stimulated intense and fruitful research in the field of chemically modified electrodes. Functional substances are immobilized on base electrodes through covalent bonds like ester, ether and amide with desired functional substances. When noble metal electrodes like gold and platinum are immersed in a solution of thiol or disulfide for a while, the electrode surface is readily modified with a self-assembled thiol monolayer. Since various functional groups can be immobilized on the electrode surface and the method of modification is versatile, this method is widely used. However, only a few examples of its application to electrosynthesis are known [50].

Pinson, Saveant and co-workers developed an efficient and general procedure for covalent bonding of aryl groups to an electrode by cathodic reduction of an aryldiazonium ion. The aryl radicals thus generated bind efficiently to the electrode surface. The resulting modified electrode is highly stable. Thus, various aryl groups can be attached to carbon, silicon and metal electrode surfaces by this procedure [51a]. Such modified electrodes are widely applicable, for instance in sensors and electrocatalysts, but application to electrosynthesis has been limited so far [51b].

5.4.4 Polymer-Modified (Coated) Electrodes

Polymer-modified (coated) electrodes have an advantage that a large number of functional substances, such as optically active compounds and electroactive compounds like redox catalysts, can be incorporated in the polymer matrix coated on the electrode surface, as shown in Figure 5.10 [45].

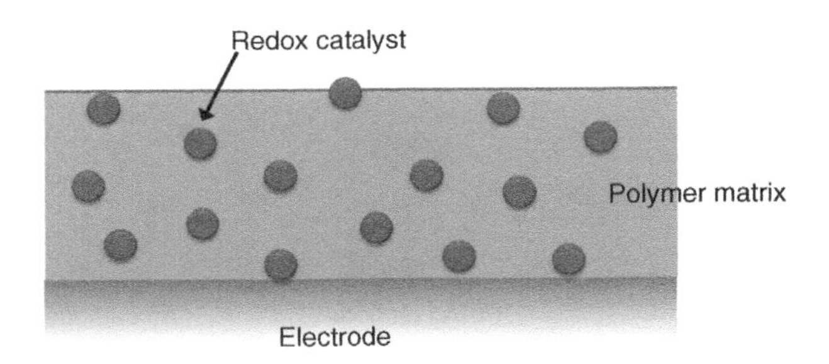

Figure 5.10 Polymer-modified electrode with redox catalyst

The polymer used for coating must be soluble in some solvent and insoluble in the solvent containing the supporting electrolyte used for electroanalytical studies and/or electrolytic reactions. Modification methods are shown below: simple dip coating and subsequent drying, polymer coating directly from monomers by chemical or electrochemical initiated polymerization, conducting polymer coating and a confirmed chemical binding like an amido bonding of the polymer layer using electrode surface anchor groups.

The following polymer materials are used for the modification of electrodes:

1. redox polymer (polyviologene, polyvinylferrocene, etc.)
2. polymers (poly-4-vinylpyridine, polyvinylsulfonic acid, polyacrylonitrile, etc.) that can form metal complexes and immobilize redox active species through columbic interaction
3. polymers (polycarbonate, polyurethane, etc.) that enable selective ion/gas permeation
4. conducting polymers.

Although examples of inorganic polymer-coated electrodes are limited, Prussian blue, heteropolyacid, clay film etc. can be utilized as modification substances. The polymer-modified electrodes have wide applications for fuel cells, capacitors, display materials, biosensors, ion/gas sensors and asymmetric synthesis, and therefore they are currently being intensively studied.

5.5 PAIRED ELECTROSYNTHESIS

Electrosynthesis involves the simultaneous occurrence of both reduction at a cathode and oxidation at an anode. In a typical inorganic electrolysis of an aqueous solution of sodium chloride, sodium hydroxide is produced at the cathode, while chlorine gas is formed at the anode. Thus, both electrode reactions are efficiently utilized. In organic electrolysis, the synthesis of desired products is performed by either cathodic or anodic reaction. Hence, in most cases, the reaction products at the counter electrode are wasted, therefore the simultaneously synchronous utilization of both cathode and anode reactions for efficient synthesis of desired products is preferable from a practical point of view. Paired electrosynthesis is based on this methodological concept. Baizer first formulated and demonstrated approaches to the above concept, and introduced the term of paired electrosynthesis [19,52]. At most 200% current efficiency could be expected in the ideal case of paired electrosynthesis using cathodic and anodic processes to provide the same product. Recently, BASF Co. in Germany commercialized paired electrosynthesis for the first time. Phthalide and *t*-butylbenzaldehyde dimethylacetal are produced simultaneously in an undivided cell. The details are described in Chapter 8.

Paired electrosynthesis is classified in the following ways based on reaction modes [19,52].

1. **Parallel paired electrosynthesis (electrosynthesis of different products from different starting substrates):** A typical example is electrosynthesis of dihydrophthalic acid and acetylene dicarboxylic acid from phthalic acid and butynediol at a cathode and anode, respectively [52]
2. **Divergent paired electrosynthesis (electrosynthesis of different products from the same substrate):** Details and an example are described in Chapter 7 (Figure 7.8).
3. **Convergent paired electrosynthesis:** This is the electrosynthesis of a single product from different starting substrates, for instance electrosynthesis of propylene oxide via chlorohydrine from propylene and chloride ion as well as electrosynthesis of sulfeneimines from amine and disulfide using bromide ion mediator [53]. As shown in

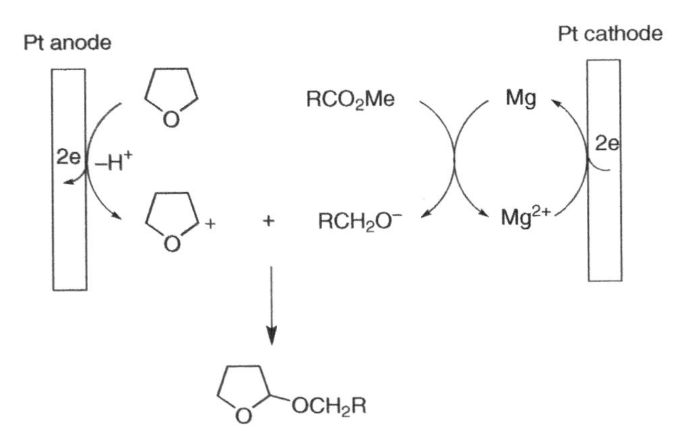

Figure 5.11 Convergent paired electrosynthesis (a)

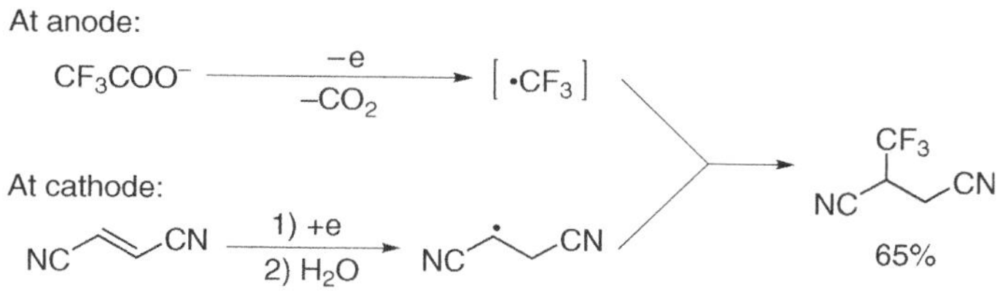

At anode:

$$CF_3COO^- \xrightarrow[-CO_2]{-e} [\cdot CF_3]$$

At cathode:

NC–CH=CH–CN $\xrightarrow[2)\ H_2O]{1)\ +e}$ NC–ĊH–CH₂–CN → NC–CH(CF₃)–CH₂–CN 65%

Figure 5.12 Convergent paired electrosynthesis (b)

Figure 5.11, aliphatic ester is reduced with reactive Mg metal cathodically derived from Mg ion and alkoxide ion is formed. On the other hand, tetrahydrofuran as a solvent is oxidized at the anode to generate the corresponding cation intermediate, which reacts with the cathodically formed alkoxide ion to give α -alkoxytetrahydrofuran [54]. In this reaction, both anodic and cathode reactions participate to produce a single product, thereby the total current efficiencies do not exceed 100%. The electrolytic reaction shown in Figure 5.12 is another example [55]. The trifluoromethylation of fumaronitrile requires both anodic and cathodic reactions, hence this reaction is a kind of paired electrosynthesis.

4. **Linear paired electrosynthesis (electrosynthesis of single product through anodic oxidation followed by cathodic reduction or vice versa):** As shown in Eq. 5.17, oxidation of 2,3-butandiol provides acetoin, and afterwards acetoin is reduced at an Hg/Zn cathode to ethyl methyl ketone [56]. The use of an undivided cell is essential in this case.

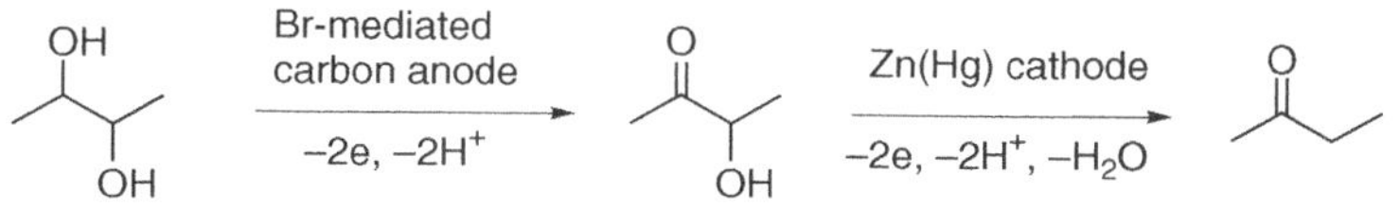

$$\text{2,3-butanediol} \xrightarrow[-2e,\ -2H^+]{\text{Br-mediated carbon anode}} \text{acetoin} \xrightarrow[-2e,\ -2H^+,\ -H_2O]{\text{Zn(Hg) cathode}} \text{ethyl methyl ketone} \quad (5.17)$$

5. **Paired electrosynthesis using EGB regenerated at the anode and cathode successively or vice versa:** Figure 5.13 illustrates an example of this kind of paired electrosynthesis. An azobenzene is reduced to give the corresponding dianion (EGB), which then reacts with a mixture of ethyl acetate and butyl bromide as the starting substrate

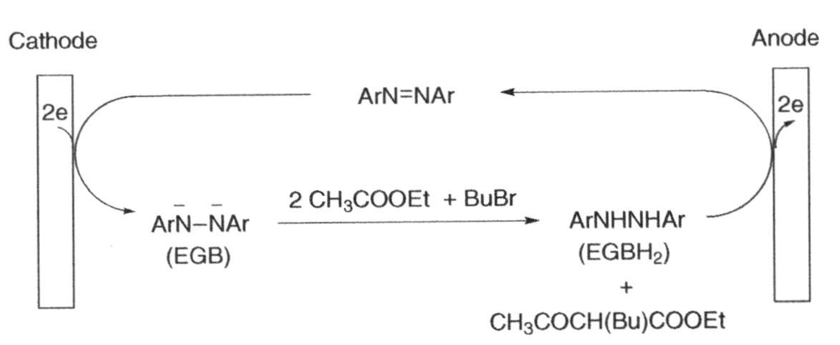

Figure 5.13 Paired electrosynthesis using regenerated EGB

to give the final product (ethyl α-(acetyl)hexanoate) and $EGBH_2$. The $EGBH_2$ is oxidized at the anode to regenerate the EGB [57].

6. **Paired electrosynthesis using different mediators regenerated at the anode and cathode:** Dioxygen (O_2) is cathodically reduced in aqueous and non-aqueous solutions to reactive oxygen species such as H_2O_2 and $O_2^{-\bullet}$, respectively, which are oxidizing reagents for organic compounds. Hence, cathodic oxidation of organic substrates is realized using such cathodically generated oxygen species. However, the concentration of cathodically generated H_2O_2 is so low that the oxidation rate of organic compounds with H_2O_2 is not enough for a practical electrolytic process. Catalysts are therefore necessary for the practical cathodic oxidation to be paired with the anodic oxidation. Nonaka and his co-workers developed a unique cathodic oxidation system using a tungstate/pertungstate redox mediator and applied this to the paired electrosynthesis, as shown in Figure 5.14 [58]. *N*-Hydroxyamine derivative as a single starting substrate is indirectly oxidized to the nitrone as a single product with HWO_5^- derived from HWO_4^- and H_2O_2 in a cathodic chamber, and with halogen (X_2) anodically formed from X^- in an anodic chamber. The total cathodic and anodic current efficiencies for the nitrone were very high (190%).

In addition, paired electrosynthesis was also developed by the combination of anodic oxidation and cathodic reduction using a hydroxyl radical derived from H_2O_2 and various metal ion redox mediatory catalysts, such as Fe^{2+}/Fe^{3+}, V^{4+}/V^{3+}, Cu^{+}/Cu^{2+} and Ni^{2+}/Ni^{3+}.

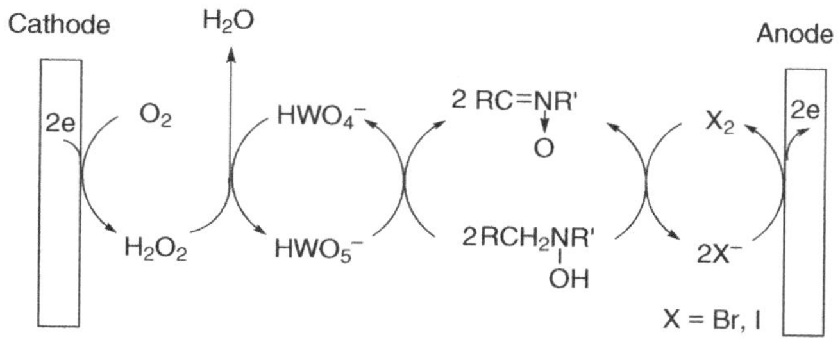

Figure 5.14 Paired electrosynthesis using cathodically generated hydrogen peroxide as an oxidant

7. **Parallel paired electrochemical polymer reaction:** Quite recently, Inagi, Fuchigami and co-workers achieved the first paired electrochemical polymer reactions of conducting polymers, such as alternating copolymers of 9-fluorenol and 9,9-dioctylfluorene adsorbed on an anode and cathode to form a 9-fluorenone moiety at the anode and a fluorene moiety at the cathode, respectively [59]. Since the reactions are solid phase, product isolation is very easy.

5.6 REACTIVE ELECTRODES

In order to avoid the decomposition of product or intermediate once it has been formed at the counter electrode, a divided cell is required. However, the divided cell increases the cell voltage (the voltage between anode and cathode) and maintenance is also troublesome. When the desired reaction is cathodic reduction, anodically dissolving metal electrodes are used as anodes to avoid both the decomposition of product or intermediate at the anode and increase the cell voltage. Such electrodes are called sacrificial electrodes, sacrificial anodes or sacrificial metal anodes [60]. For instance, as shown in Eq. 5.18, cathodic carboxylation of organic halides using a magnesium anode provides carboxylic acid selectively in high yield [61]. In this case, the carboxylate anion formed is trapped with magnesium ion to precipitate. Both the anodic oxidation of the formed carboxylate ion and its esterification with unreacted organic halide can therefore be avoided. In such cathodic carboxylation, aluminium and zinc anodes, as well as magnesium anodes, are often used.

Cathode:

$$R{-}X + 2e \longrightarrow R^- + X^-$$

$$R^- + CO_2 \longrightarrow RCOO^-$$

Anode:

$$Mg \longrightarrow Mg^{2+} + 2e \qquad (5.18)$$

Total:

$$R{-}X + CO_2 + Mg \longrightarrow RCOOMgX \text{ or } (RCOO)_2Mg + X^-$$

Recently, a sacrificial metal anode was used in electrosynthesis. The resulting metal ions derived from the anode, e.g. Mg, Zn, Al and, Cu, often play important roles in electrochemical reactions. These metal ions and cathodically generated reactive species form highly reactive intermediates new reagents or trap halide ions generated by cathodic reduction of silyl chlorides to polysilanes (see section 5.8.6) [62]. Moreover, such anodically dissolving metal ions significantly affect regiochemistry, stereochemistry and product selectivity by their coordination and catalytic effects as well as formation of a new reagents *in situ* [63]. For instance, as shown in Figure 5.15, zinc ion generated from the anode and trifluoromethyl anion cathodically derived from CF_3Br forms an organometallic compound as an intermediate, which undergoes a Reformatsky type reaction with aldehydes and ketones to provide trifluoromethylated alcohols. In contrast, the cathodic reduction of CF_3Br in the presence of zinc ions does not provide any desired product [64]. Hence, anodically

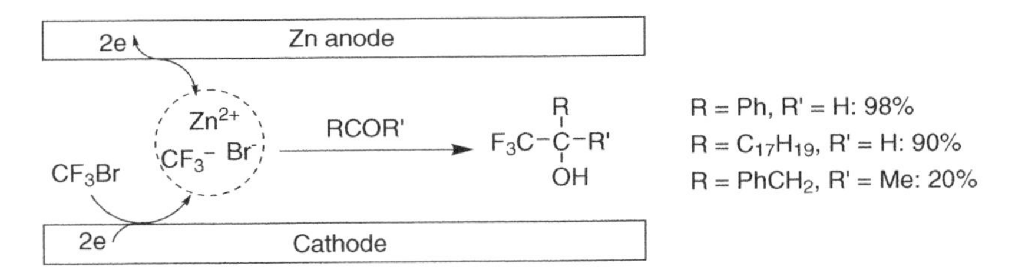

Figure 5.15 Electrosynthesis using reactive electrode

generated zinc ions are very specific and quite different from ordinary zinc ions.

Other typical examples are the cross-coupling reaction of activated halogenated compounds and carbonyl compounds, and the cross-coupling reaction between non-halogenated compounds [65].

5.7 ELECTROCHEMICAL FLUORINATION

Organofluorine compounds are classified into two groups: perfluoro compounds and partially fluorinated compounds. The compounds in the former class are widely utilized as functional materials, while those in the latter family find biological uses as pharmaceuticals and agrochemicals. Perfluoro compounds are manufactured by converting all C–H bonds to C–F bonds using electrochemical fluorination in anhydrous liquid HF as a solvent with a nickel anode. This process is called electrochemical perfluorination and has already been commercialized (see Chapter 8) [66–69]. Electrochemical perfluorination in KF-2HF melts at a carbon anode has also been developed for the preparation of perfluorinated low-molecular-weight organic compounds [69].

Electrochemical partial fluorination, or selective electrochemical fluorination, is a new method [69–72]. Since the discharge potential of the fluoride ion is extremely high (>+2.9 V vs. SCE at Pt anode in MeCN), the fluorination proceeds via a (radical) cation intermediate, as shown in Eq. 5.19, which is the general pathway for anodic nucleophilic substitutions.

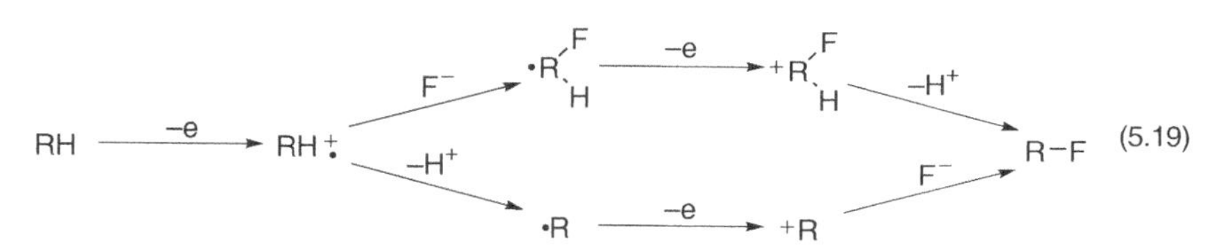

In this method, fluorine gas is not generated and no hazardous reagents are required, therefore this electrochemical fluorination is much safer than the conventional chemical method, which often requires hazardous and/or difficult-to-handle reagents.

Selective electrochemical fluorination can commonly be achieved in aprotic solvents such as acetonitrile (MeCN), dichloromethane, dimethoxyethane (DME), nitromethane and sulfolane containing fluoride ions to provide mostly mono- and/or difluorinated products [72]. Electrolyses are conducted at constant potentials slightly higher than the first oxidation potential of the substrate by using a platinum or graphite anode. Constant current electrolysis is also effective for selective fluorination in many cases. The choice of the combination of a supporting fluoride salt and an electrolytic solvent is important to accomplish efficient selective fluorination because competitive anode passivation (the formation of a non-conducting polymer film on the anode surface that suppresses the Faradaic current) takes place very often during the electrolysis. Pulse electrolysis is in many cases effective in order to avoid such passivation, therefore difficult-to-oxidize fluoride salts, which do not cause the passivation of the anode and have strongly nucleophilic F^-, are generally recommended as the supporting fluoride salts. Thus, room temperature molten salts such as R_3N-*n*HF ($n=3$–5), R_4NF-*n*HF ($n=3$–5) and pyridine poly(hydrogen fluoride) salt (Py-*n*HF) are most often used and even R_4NBF_4 and R_4NPF_6 salts are effective in some cases [70–72]. Particularly when HF supporting salts and low hydrogen overpotential cathodes such as platinum are used, the reduction of protons (hydrogen evolution) occurs predominantly at the cathode during the electrolysis. A divided cell is therefore not always necessary for fluorination under such conditions.

In aprotic solvent, F^- becomes more nucleophilic, but the reactivity of F^- is quite sensitive to the water content of the electrolysis system because a hydrated F^- is a weak nucleophile. Drying of both the solvent and electrolyte is therefore necessary to optimize the formation of fluorinated products.

A few examples are given below, and more examples have been reported in the literature.

5.7.1 Electrochemical Fluorination of Aromatic Rings

Aromatic compounds such as benzene, substituted benzenes and naphthalene are selectively fluorinated by constant potential anodic oxidation in Et_4N-3HF/MeCN or Et_4NF-4HF without solvent (Eq. 5.20) [68,69]. Fluorination proceeds via addition with two fluoride ions followed by elimination of HF to provide monofluorinated aromatic compounds. Notably, *ipso*-substitution with fluorine also takes place in high yield, as shown in Eq. 5.21.

$$\text{naphthalene} \xrightarrow[\text{Et}_4\text{NF-3HF/MeCN, 1.8 V vs. SCE}]{-2e,\ -H^+} \text{1-fluoronaphthalene (27\%)} + \text{1,4-difluoronaphthalene (3\%)} \tag{5.20}$$

$$\text{1,4-di-}tert\text{-butylbenzene} \xrightarrow[\text{Et}_4\text{NF/MeCN}]{-2e} \text{1-}tert\text{-butyl-4-fluorobenzene (80\%)} \tag{5.21}$$

5.7.2 Electrochemical Fluorination of Olefins

Electrochemical fluorination of olefins provides mono- and/or difluorinated products. The selectivity depends on the molecular structure of the substrate and the electrolytic solvent, as shown in Eq. 5.22 [72,73]. For α-acetoxystyrene and 1-acetoxy-3,4-dihydronaphthalene, the corresponding α-fluoroketones are formed as shown in Eq. 5.23 [72,74].

$$R^1CH{=}CHR^2 \xrightarrow[\text{F}^-\text{/MeCN or AcOH}]{-2e} R^1CH(F){-}CH(Y)R^2 \quad Y = \text{F, NHCOMe, OAc} \tag{5.22}$$

$$Ar(AcO)C{=}CHR^1 \xrightarrow[\text{Et}_3\text{N-3HF/MeCN}]{-2e,\ -Ac^+} ArC(=O){-}CH(F)R^1 \quad \text{44–63\%} \tag{5.23}$$

5.7.3 Benzylic Electrochemical Fluorination

Although anodic benzylic substitution reactions take place readily, anodic benzylic fluorination does not always occur. The major competitive reaction is acetamidation when MeCN is used as a solvent. For example, electrochemical benzylic fluorination in MeCN proceeded selectively when the *p*-position of the phenyl group was substituted by an electron-donating group, as shown in Eq. 5.24 [75]. In contrast, α-acetoamidation became the major reaction when the phenyl group had no electron-donating substituent.

$$\text{X-C}_6\text{H}_4\text{-CH}_2\text{COMe} \xrightarrow[\text{F}^-/\text{MeCN}]{-2e,\ -\text{H}^+} \text{X-C}_6\text{H}_4\text{-CHF-COMe} + \text{X-C}_6\text{H}_4\text{-CH(NHCOMe)-COMe} \quad (5.24)$$

X = H	7%	34%
X = MeO	69%	< 1%

Mono- and difluorination can be performed selectively depending on applied potential.

5.7.4 Electrochemical Fluorination of Sulfides

Fuchigami and co-workers found that anodic fluorination of sulfides having α-electron-withdrawing groups proceeded quite well to provide the corresponding α-fluorinated products in good yields, as shown in Eq. 5.25 [72,76]. The fluorination proceeds by way of a Pummerer-type mechanism via the fluorosulfonium cation (**A**), as shown in Eq. 5.26 [77,78]. Thus, when R is an electron-withdrawing group, the deprotonation of **A** is significantly facilitated, and consequently the fluorination proceeds efficiently.

$$\text{R}^1\text{S-CH}_2\text{-R}^2 \xrightarrow[\text{F}^-/\text{MeCN or DME}]{-2e,\ -\text{H}^+} \text{R}^1\text{S-CHF-R}^2 \quad \sim 98\% \quad (5.25)$$

R^1 = Ar, Alkyl, Bzl, 2-Thiazolyl, 2-Pyridyl, 2-Pyrimidyl
R^2 = CF_3, CN, COR, COOR, $CONR_2$, $PO(OEt)_2$, C≡CH

$$\text{PhS-CH}_2\text{R} \xrightarrow[\text{F}^-]{-2e} \underset{\textbf{A}}{\text{Ph}\overset{+}{\text{S}}(\text{F})\text{-CHR(H)}} \xrightarrow[-\text{HF}]{} \text{Ph}\overset{+}{\text{S}}\text{=CHR} \xrightarrow{\text{F}^-} \text{PhS-CHFR} \quad (5.26)$$

5.7.5 Electrochemical Fluorination of Heterocyclic Compounds

Many heterocyclic compounds have specific biological activities, while it is known that introduction of fluorine atom(s) to organic molecules dramatically changes or enhances their biological activities. However, selective fluorination of heterocyclic compounds using conventional fluorinating reagents is not easy. Many successful examples of selective anodic fluorination of heterocycles containing sulfur, nitrogen, oxygen and phosphine have been reported to date [72,79]. Some examples are shown in Table 5.4.

The selectivity of fluorination is also strongly influenced by supporting fluoride salts, as shown in Eq. 5.27 [80]. Since Et_3N-3HF contains the free base Et_3N, the difluorinated product, once formed, is dehydrofluorinated to the monofluoro product.

−2e, Et_4NF-4HF/MeCN: 68% (*cis/trans* = 2)

−2e, Et_3N-3HF/MeCN: 58%

(5.27)

Furthermore, anodic difluorination accompanied by C−C double bond cleavage is also known, as shown in Eq. 5.28 [81].

−4e, Et_3N-3HF/DME → 2

Ar = Ph : 67%
= *p*-BrC_6H_4 : 62%

(5.28)

Anodic fluorination of heterocyclic compounds derived from optically active α-amino acids proceeds in good yields and with high diastereoselectivity, as shown in Eqs. 5.29 and 5.30. Notably, fluorination does not occur at the benzylic position and fluorination predominantly takes place α to the carbonyl group, as shown in Eq. 5.31 [82].

Table 5.4 Electrochemical fluorination of heterocyclic compounds

Substrate	Salt/solvent	Product	Yield (%)
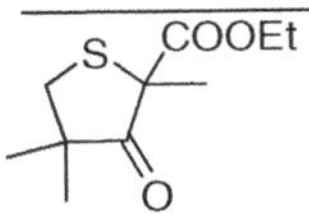 S, COOEt, O	Et_3N-3HF/MeCN	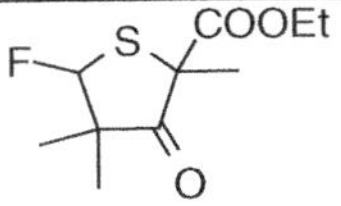 F, S, COOEt, O	74 (*trans*/*cis* = 74/26)
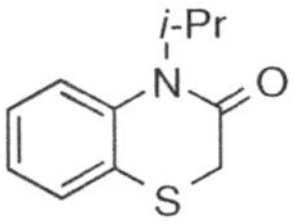 *i*-Pr, N, O, S	Et_3N-3HF/MeCN	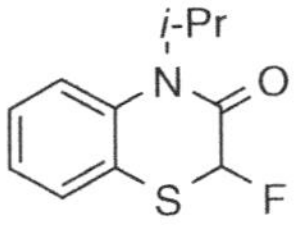 *i*-Pr, N, O, S, F	88
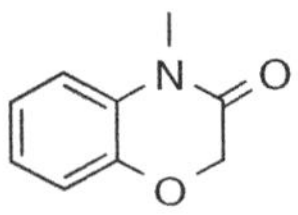 N, O, O	Et_4NF-4HF/DME	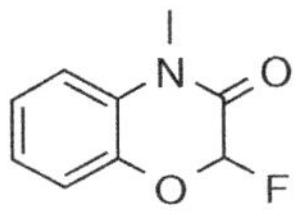 N, O, O, F	66
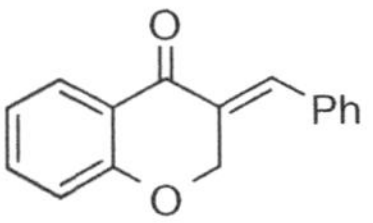 O, Ph, O	Et_4NF-4HF/DME	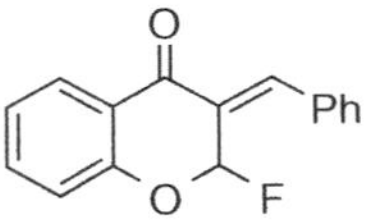 O, Ph, O, F	72
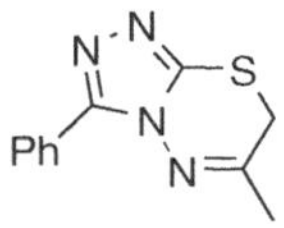 N, N, S, N, N, Ph	Et_4NF-4HF/DME	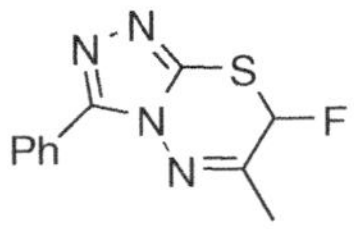 N, N, S, F, N, N, Ph	81

COOMe, S, NCOR → F, COOMe, S, NCOR

$$\xrightarrow[\text{Et}_3\text{N-4HF/DME}]{-2e,\ -H^+} \tag{5.29}$$

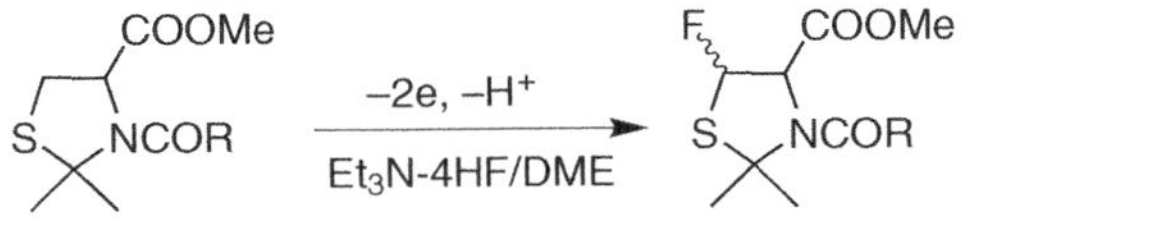

R = *p*-Tol, Ph, Me, H

52–91% yield, 59–95% de

Me, COOMe; O, NCOPh $\xrightarrow[\text{Et}_4\text{NF-4HF/MeCN}]{-2e,\ -H^+}$ Me, F, COOMe; O, NCOPh (5.30)

73% yield, 81% de

S, R, X, O $\xrightarrow[\text{Et}_3\text{N-3HF/MeCN}]{-2e,\ -H^+}$ F, S, R, X, O (5.31)

(*trans* major)

R = Ph, 2-Naphthyl,Mesityl,Et,n-Pr

X = NH,MeN, *i*-PrN, PhN, BzlN,O,S

Yield:58~86%

5.7.6 Electrochemical Fluorination of Heterocyclic Compounds with PhS Group as Electroauxiliary

Heterocyclic compounds having a phenylsulfenyl group are also anodically fluorinated in good yields, as shown in Eqs. 5.32 and 5.33 [83,84].

PhS, O, N, *n*-Bu $\xrightarrow[\text{Et}_3\text{N-3HF/MeCN}]{-2e,\ -H^+}$ F, PhS, O, N, *n*-Bu (5.32)

92%

ArS, O, P–O, O $\xrightarrow[\text{Et}_4\text{NF-3HF/DME}]{-2e,\ -H^+}$ F, SAr, O, P–O, O (5.33)

58%

It is noted that fluorination product selectivity is dramatically changed depending on electrolytic solvents, as shown in Eq. 5.34 [85].

p-ClC$_6$H$_4$S, F, O, O, O $\xleftarrow[\text{Et}_4\text{NF-4HF/DME}]{-2e,\ -H^+}$ *p*-ClC$_6$H$_4$S, O, O, O $\xrightarrow[\text{Et}_4\text{NF-5HF/CH}_2\text{Cl}_2]{-e}$ F, O, O, O

80% 96%

(5.34)

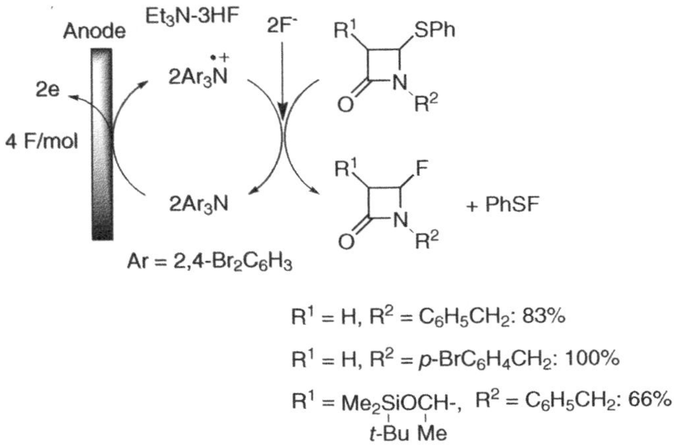

Anode $Et_3N\text{-}3HF$ $2F^-$

2e $2Ar_3N^{\bullet+}$ R^1 SPh N R^2 O

4 F/mol

$2Ar_3N$ R^1 F N R^2 O + PhSF

Ar = 2,4-$Br_2C_6H_3$

$R^1 = H$, $R^2 = C_6H_5CH_2$: 83%

$R^1 = H$, $R^2 = p\text{-}BrC_6H_4CH_2$: 100%

$R^1 = Me_2SiOCH-$, $R^2 = C_6H_5CH_2$: 66%
t-Bu Me

Figure 5.16 Electrocatalytic fluorodesulfurization of β-lactams

As mentioned earlier, anode passivation takes place often, resulting in poor yield and low current efficiency. In order to avoid such passivation, various mediators, such as halide ions, triarylamines and iodoarenes, can be used, as shown in Figure 5.16 [86].

Moreover, ethereal solvents such as DME are much more suitable than MeCN for the anodic fluorination of various heterocyclic sulfides (Eq. 5.35) [87]. The pronounced solvent effect of DME could be explained in terms of the significantly enhanced nucleophilicity of fluoride ions as well as the suppression of anode passivation and overoxidation of fluorinated products.

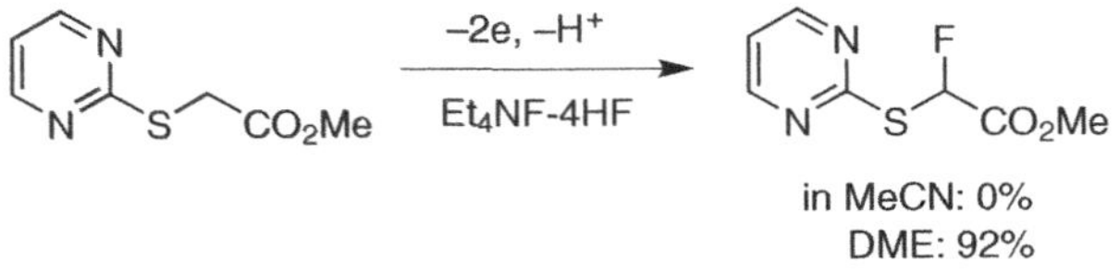

N N S CO_2Me $\xrightarrow[Et_4NF\text{-}4HF]{-2e,\ -H^+}$ N N S F CO_2Me (5.35)

in MeCN: 0%
DME: 92%

5.7.7 Electrochemical Fluorination Using Inorganic Fluoride Salts

Inorganic fluoride salts such as alkali-metal fluorides (MFs) are stable, easy to handle and inexpensive. They are therefore strong candidates for

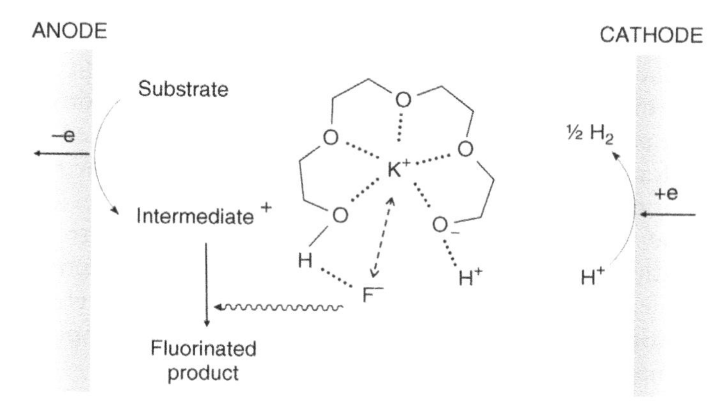

Figure 5.17 Poly(ethylene glycol) with two terminal hydroxy groups serves as a multifunctional additive for anodic fluorination using KF

reagents in nucleophilic fluorination as well as supporting electrolytes in chemical and electrochemical fluorination. The challenge is to overcome problems such as poor solubility and low nucleophilicity of MFs in organic solvents. Phase-transfer catalysts such as crown ethers and quaternary ammonium or phosphonium salts are known to reduce the coulombic interactions of MFs and are commonly used for this purpose. Fuchihgami and co-workers reported successful anodic fluorination in combination with an electrochemical method using a poly(ethylene glycol)/MF system where the MF is either KF or CsF, as shown in Figure 5.17 [88].

5.8 ELECTROCHEMICAL POLYMERIZATION

Electrochemical polymerization uses electrogenerated substrates for polymerization as monomers or initiator. In the former case, the electrogenerated aromatic monomer couples in a polycondensation process to give a π-conjugated polymer, in which the *p*-orbitals of the aromatics overlap throughout the polymer main chain. Such conjugated polymers are intrinsically conductive and therefore are called conducting polymers [89]. Generally they are semi-conducting materials but chemical or electrochemical doping imparts electrical conductivity to them. The conducting polymers are obtained on the surface of the working electrode as films because the electron transfer of monomer and its coupling reaction proceeds near the surface of the electrode and insoluble polymeric product is deposited on it. Although chemical electron transfer of monomers in solution is also available to produce the corresponding conducting polymers, the insoluble polymer powder obtained is difficult to process for application. The easy film formation during electrochemical polymerization is useful for application in electrochemical devices such as sensors and displays [90]. This chapter deals with electro-oxidative polymerization, electro-reductive polymerization and electrosynthesis of polysilane. Polymerization of vinyl monomers using an electrogenerated initiator system is summarized elsewhere [91], although recent developments in this process are included in this chapter.

5.8.1 Electro-oxidative Polymerization of Aromatic Monomers

The electron-rich aromatic and hetero-aromatic monomers listed in Figure 5.18 can be easily oxidized on the anode surface to form their radical cation states. The radical cations couple to form C–C bonds, and following deprotonation this leads to a neutral dimer (Eq. 5.36). The generated conjugated dimer has a lower oxidation potential than that of the monomer, so further oxidation of the dimer leads to oligomerization and polymerization. The highly conjugated polymers are no longer soluble in electrolytic medium and therefore are deposited on the anode surface as a film. The deposited polymer is oxidatively doped during the application of potential for polymerization, thus the film formed is not passive but still conductive to carry out a continuous electrochemical reaction on the electrode. In this procedure, the coupling position of monomer can be predicted by the spin-density distribution on the aromatic ring [92]. The introduction of electro-auxiliaries such as a silyl group can enable dominant coupling at a specific position [93].

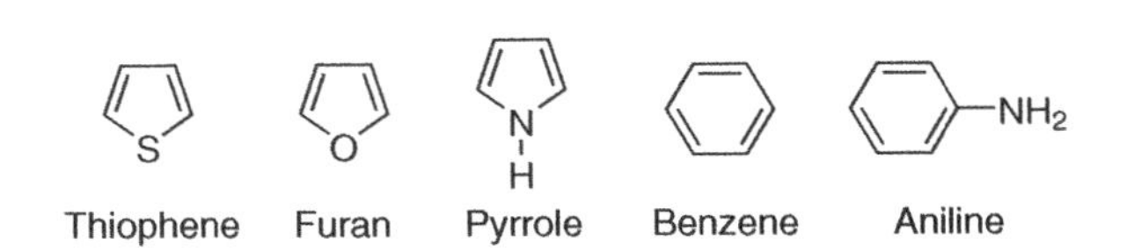

Figure 5.18 Aromatic monomers for oxidative polymerization

X = O, S, NR

(5.36)

5.8.2 Electrochemical Polymerization

There are several methods of electrochemical polymerization, for example the potential sweep, potentiostatic (constant-potential) and galvanostatic (constant-current) methods. A suitable method should be chosen depending on the monomers and conditions used. Cyclic voltammetry (CV) measurement of a monomer in a three-electrode setup gives information on the oxidation potential for electrochemical polymerization. To prevent overoxidation of a product polymer, the application of a higher potential to a working electrode for a long period should be avoided since the product polymer with extended π-conjugation is easily oxidized compared to the corresponding monomer. The potential sweep and potentiostatic methods require milder polymerization conditions than the galvanostatic method.

In the potential sweep method, polymerization is carried out with CV analyzer. A working electrode, a counter electrode and a reference electrode are put together in an electrolytic cell containing electrolyte and monomer. The repeating potential sweep across the potential range determined from the advanced CV measurement produces product polymer film on the working electrode and this electron-transfer process can be monitored by CV, as shown in Figure 5.19, which shows typical cyclic voltammograms of pyrrole during electropolymerization. The first scan represents the oxidation current of pyrrole monomer and then polymerization proceeds at the surface of the working electrode, resulting in an irreversible voltammogram. In the course of deposition of the product polypyrrole during repeated scans, anodic and cathodic currents derived from redox of polypyrrole at a lower potential range gradually increase.

In the potentiostatic method a three-electrode system is used and the applied potential to the working electrode is controlled by a potentiostat

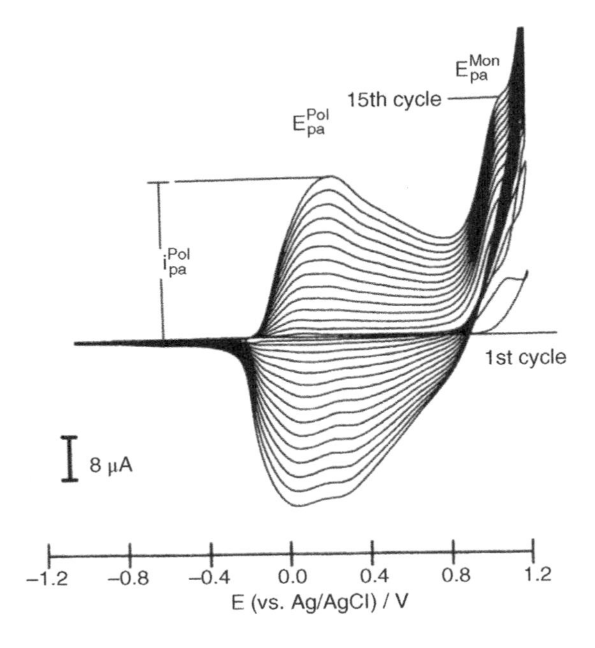

Figure 5.19 Typical cyclic voltammograms of pyrrole monomer

with a monitoring charge. In the galvanostatic method a two-electrode system consisting of a working electrode and a counter electrode is used, and constant current is passed between them by a DC power source. This process cannot control potential on the working electrode, which may cause problems such as overoxidation of the product polymers.

5.8.3 Conditions for Electrochemical Polymerization

The concentration of monomer is one important factor in electrochemical polymerization. If the polymerization does not proceed at the electrode surface, the use of a lot of monomer may be effective to promote film formation of the conducting polymers. The choice of material for the working electrode is also important in terms of not only overpotential but also compatibility with conducting polymer films.

In addition to conventional electrolytic solutions such as the supporting salt/solvent, room temperature ionic liquids are known to be suitable as polymerization media [94]. There are several advantages of electropolymerization in ionic liquids: (i) electropolymerization proceeds rapidly,

(ii) the compatibility of the product film with the electrode surface is good and (iii) product polymers have a dense and smooth surface, high electric conductivity, high electrochemical capacitance and good reversibility of redox cycles.

Recently boron trifluoride–ether complex has been shown to be attractive as a polymerization medium [95]. Boron trifluoride acts as a Lewis acid to form a complex with aromatic monomer, which reduces the oxidation potential of the monomer. This method is therefore effective for polymerization of aromatic monomers with relatively high oxidation potentials. The reduction in polymerization potential can avoid overoxidation of product polymers.

5.8.4 Electrochemical Doping

If a polymer film on a working electrode is obtained by electropolymerization, it should be put into electrolytic solution without monomer (monomer-free electrolyte) in order to study the redox behaviour of the conducting polymer. Typically a pair of broad redox responses is observed in the voltammogram. The injection or removal of electrons to or from conducting polymers results in the formation of polarons and bipolarons in the repeating structure (commonly known as doping), producing considerable variations in the physical properties of the polymer itself and imparting features such as drastic colour changes and electrical conductivity [89]. The charges thus generated along the polymer must be compensated for by the addition of neighbouring ions (or dopants) and the insertion and release of such dopants may induce volume changes in the conjugated polymer. The doping ratio can be estimated from the amount of charge generated in the conducting polymers.

5.8.5 Electro-reductive Polymerization of Aromatic Monomers

Aromatic dihalides are representative monomers for electroreductive polymerization. Cathodic reduction of an aromatic dihalide generates a radical anion, followed by polymerization accompanying elimination of halide (Eq. 5.37). Bond formation takes place at the specific position of the monomer where the halogen atom is substituted.

$$\mathrm{Br{-}Ar{-}Br} \xrightarrow{+2e,\ -2Br^-} \left(\mathrm{Ar}\right)_n \tag{5.37}$$

On electroreductive polymerization of 1,4-dibromopyridine, addition of a catalytic amount of nickel(II) complex is effective and the complex works as a mediator. Electrogenerated nickel(0) promotes dehalogenation of the dibromopyridine followed by its polymerization [96]. Electro-oxidative polymerization of pyridine is difficult because of its high oxidation potential, but this reductive polymerization enables polypyridines to be produced.

Poly(*p*-phenylene vinylene) is also synthesized by electroreductive polymerization. As shown in Eq. 5.38, the electrochemical reduction of α,α,α′,α′-tetrabromo-*p*-xylene generates a quinodimethane intermediate and its subsequent polymerization gives a poly(xylylene). This is further reduced to poly(*p*-phenylene vinylene) [97].

Br Br / Br Br $\xrightarrow{+2e,\ -2Br^-}$ [Br / Br] $\longrightarrow$ Br / Br, n $\xrightarrow{+2ne,\ -2Br^-}$ n

(5.38)

5.8.6 Applications of Conducting Polymers

The physical properties of conducting polymers are drastically changed by the doping process. Table 5.5 summarizes the application of conducting polymers. The undoped (neutral) state of conducting polymers is generally semi-conducting, and it can be applied to material for transistor, photovoltaic cell and electroluminescence devices. On the other hand, electrical

Table 5.5 Applications of conducting polymer films

Property	Application
Conductivity (doped state)	Conducting materials, antistatics, capacitor materials
Doping/dedoping process	Electrochromic materials, sensors, actuators, memory
Semi-conductivity (undoped state)	Transistors, photovoltaic cells, electroluminescent materials

conductivity appears in the doped state of conducting polymers, which is useful in conducting materials, anti-statics and capacitor materials. The doping/dedoping process of conducting polymers is also useful for electrochromic devices, sensors and actuator applications. Application of conducting polymers for electronics will be described in Chapter 7.

5.8.7 Electrochemical Synthesis of Polysilanes

Polysilanes composed of linearly connected Si–Si bonds are a class of σ-conjugated polymers and show unique optical properties [98]. The Kipping method, a well-known synthetic method of obtaining polysilanes from a dichlorosilane, is powerful but the use of Na metal is indispensable. Electrochemical reduction of chlorotrimethylsilane affords hexamethyldisilane without the use of a reducing reagent. Although polymerization of dichlorosilanes by electrochemical reduction is rather difficult, the use of an Mg sacrificial anode results in the effective formation of polysilanes (Eq. 5.39) [99]. The mechanism proposed is that Mg^{2+},once it has been anodically dissolved from the anode, is reduced at the cathode to give reactive Mg. This involves Si–Si bond formation of monomers. Cu and Al are also available as sacrificial anode materials for reductive polymerization of polysilanes.

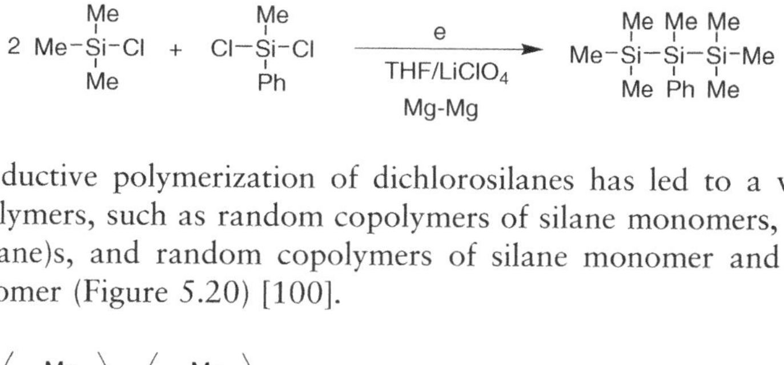

$$2\ Me_3SiCl + Cl_2SiMePh \xrightarrow[\text{THF/LiClO}_4,\ \text{Mg-Mg}]{e} Me_3Si{-}SiMePh{-}SiMe_3 \qquad (5.39)$$

Reductive polymerization of dichlorosilanes has led to a variety of copolymers, such as random copolymers of silane monomers, poly(carbosilane)s, and random copolymers of silane monomer and germane monomer (Figure 5.20) [100].

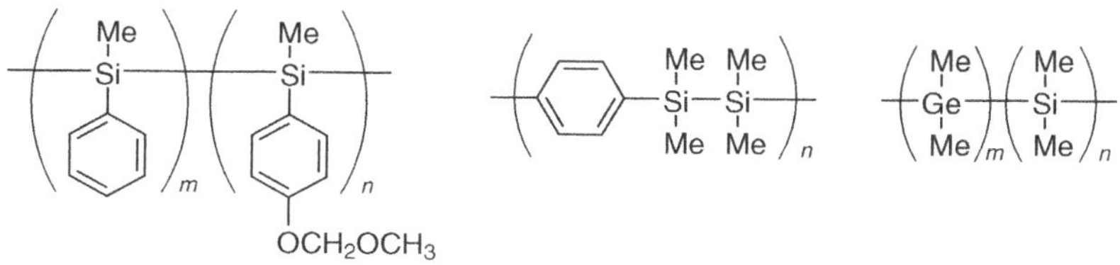

$$\left(SiMe(C_6H_5)\right)_m\left(SiMe(C_6H_4OCH_2OCH_3)\right)_n \qquad \left(C_6H_4{-}SiMe_2{-}SiMe_2\right)_n \qquad \left(GeMe_2\right)_m\left(SiMe_2\right)_n$$

Figure 5.20 Structures of copolymers based on silane and germane prepared by reductive polymerization

5.8.8 Chain Polymerization Initiated with Electrogenerated Reactive Species

Electrogenerated reactive species of small molecules can work as polymerization initiators for ionic and radical polymerization of vinyl monomers [91]. For example, a methyl radical generated by Kolbe electrolysis of acetate can be an initiator for radical polymerization of olefins. The heterogeneity of initiator concentration in the electrochemical setup makes it difficult to obtain high molecular weight polymers.

Recent progress in controlled radical polymerization using transition metal catalysts is remarkable both in academia and industry [101]. As shown in Eq. 5.40, the equilibrium (k_a/k_{da}) between domant species (A) and active species (B) controlled by redox of a transition metal catalyst can control the rate of monomer consumption, resulting in polymerization with a controlled manner. The external stimuli by electrochemical oxidation and reduction of the metal catalyst change the equilibrium and consequently more precise control of polymerization is possible (Eq. 5.40) [102].

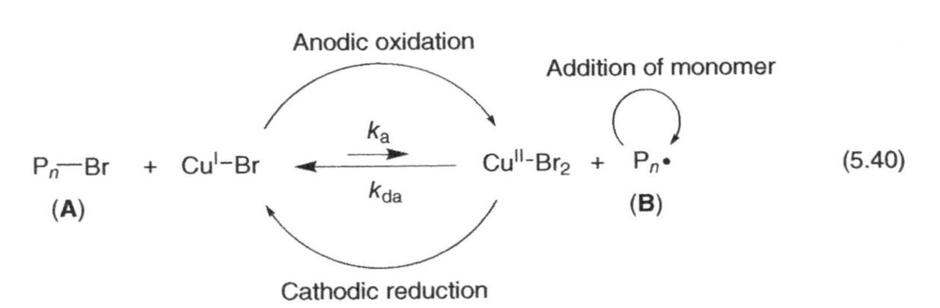

(5.40)

REFERENCES

1. (a) Hammerich, O. and Speiser, B. (eds) (2014) *Organic Electrochemistry*, 5th edn, CRC/Taylor & Francis. (b) Torii, S. (2006) *Electroorganic Reduction Synthesis*, Vols 1 and 2, Kodansha and Wiley-VCH Weinheim GmbH. (c) Bard, A.J. and Stratmann, M. (eds) (2004) *Encyclopedia of Electrochemistry*, Vol, 8 (ed. H.J. Schäefer), *Organic Electrochemistry* John Wiley & Sons. (d) Lund, H. and Hammerich, O. (eds) *Organic Electrochemistry*, 4th edn, Marcel Dekker, New York. (e) Grimshaw, J. (2000) *Electrochemical Reactions and Mechansims in Organic Chemistry*, Elsevier, Amsterdam. (f) Fry, A.J. (1989) *Synthetic Organic Electrochemistry*, Wiley Interscience, New York. (g) Shono, T. (1984) *Electroorganic Chemistry as a Tool in Organic Synthesis*, Springer-Verlag, Berlin. (h) Rifi, M.R. and Covitz, F.H. (1974)

Introduction to Organic Electrochemistry Techniques and Applications in Organic Synthesis, Marcel Dekker, New York. (i) Mann, C.K. and Barnes, K.K. (1970) *Electrochemical Reactions in Nonaqueous Systems*, Marcel Dekker, New York.

2. (a) Saveant, J.M. (1980) *Acc. Chem. Res.*, **13**, 323–329. (b) Steckhan, E. (1986) *Angew. Chem.*, **98**, 681–699. (c) Torii, S. (1986) *Synthesis*, 873–886. (d) Lund, H. and Hammerich, O. (eds) (2001) *Organic Electrochemistry*, 4th edn, Marcel Dekker, New York, Chapter 29. (e) Steckhan, E. (1987) *Topics in Current Chemistry 142. Electrochemistry I*, Springer-Verlag, Berlin, pp. 1–69.
3. Hosoi, K., Inagi, S., Kubo, T. and Fuchigami, T. (2011) *Chem. Commun.*, **47**, 8632–8634.
4. Francke, R. and Little, R.D. (2014) *J. Am. Chem. Soc.*, **136**, 427–435.
5. Torii, S., Uneyama, K., Tanaka, H., Yamanaka, Y., Yasuda, T., Ono, M. and Kohmoto, Y. (1981) *J. Org. Chem.*, **48**, 3312–3315.
6. Torii, S., Tanaka, H., Saitoh, N., Siroi, T., Sasaoka, M. and Nokam, J. (1981) *Tetrahedron Lett.*, **22**, 3193–3196.
7. Torii, S., Tanaka, H., Saitoh, N., Siroi, T., Sasaoka, M. and Nokam, J. (1982) *Tetrahedron Lett.*, **23**, 2187–2188.
8. Torii, S., Uneyama, K., Nakai, T. and Yasuda, T. (1981) *Tetrahedron Lett.*, **22**, 2291–2294.
9. Shono, T., Matsumura, Y., Hayashi, J. and Mizoguchi, M. (1980) *Tetrahedron Lett.*, **21**, 1867–1870.
10. Fujita, T. and Fuchigami, T. (1994) *J. Org. Chem.*, **59**, 7190–7192.
11. Hara, S., Sekiguchi, M., Ohmori, A., Fukuhara, T. and Yoneda, N. (1996) *Chem. Commun.*, 1899–1900.
12. Scheffold, R., Dike, M., Dike, S., Herold, T. and Walder, L. (1980) *J. Am. Chem. Soc.*, **102**, 3642–3644.
13. Scheffold, R., Rytz, G., Walder, L. and Orlinski, R. (1983) *Pure Appl. Chem.*, **55**, 1791–1797.
14. Shimakoshi, H., Nakazato, A., Hayashi, T., Tachi, Y. Naruta, Y. and Hisaeda, Y. (2001) *J. Electroanal. Chem.*, **507**, 170–176.
15. Torii, S., Tanaka, H. and Morisaki, K. (1985) *Tetrahedron Lett.*, **26**, 1655–1658.
16. Torii, S., Tanaka, H. Hamatani, H., Morisaki, K., Jutand, A., Pfluger, F. and Fauvarque, J.F. (1986) *Chem. Lett.*, 169–170.
17. (a) Tanaka, H., Chou, J., Mine, M. and Kuroboshi, K. (2004) *Bull. Chem. Soc. Jpn.*, 77, 1745–1755. (b) Kubota, J., Ido, T., Kuroboshi, M., Tanaka, H., Uchida, T. and Shimamura, K. (2006) *Tetrahedron*, **62**, 4769–4773.
18. Tanaka, H., Kawakami, Y., Goto, K. and Kuroboshi, M. (2001) *Tetrahedron Lett.*, **42**, 445–448.
19. Baizer, M.M. (1984) *Tetrahedron*, **45**, 935–969.
20. Shono, T., Kashimura, S., Ishizaki, K. and Ishige, O. (1983) *Chem. Lett.*, 1311–1312.
21. Shono, T., Ishifune, M., Ishige, O., Uyama, H. and Kashimura, S. (1990) *Tetrahedron Lett.*, **31**, 7181–7184.
22. Shono, T., Ishifune, M., Okada, T. and Kashimura, S. (1991) *J. Org. Chem.*, **56**, 2–4.
23. Fuchigami, T. and Nakagawa, Y. (1987) *J. Org. Chem.*, **52**, 5276–5277.
24. Komori, T., Nonaka, T., and Fuchigami, T. (1986) *Chem Lett.*, 11–12.
25. Fuchigami, T., Awata, T., Nonaka, T. and Baizer, M. M. (1986) *Bull Chem Soc. Jpn.*, **59**, 2873–2879.

26. Shono, T., Kashimura, S. and Ishizaki, K. (1984) *Electrochim. Acta*, **29**, 603.
27. Awata, T., Baizer, M.M., Nonaka, T. and Fuchigami, T. (1985) *Chem Lett.*, 371–374.
28. Hashiba, S., Fuchigami, T. and Nonaka, T. (1989) *J. Org. Chem.*, **54**, 2475.
29. Aller, P.M., Hess, U., Foote, C.S. and Baizer, M.M. (1982) *Syn. Commun.*, **12**, 123–129.
30. Uneyama, K., Nishiyama, N. and Torii, S. (1984) *Tetrahedron Lett.*, **25**, 4137–4138.
31. Inokuchi, T., Tanigawa, S. and Torii, S. (1990) *J. Org. Chem.*, **55**, 3958–3961.
32. Fuchigami, T., Yamamoto, K. and Yano, H. (1992) *J. Org. Chem.*, **57**, 2946–2950.
33. (a) Nonaka, T. and Fuchigami, T. (2001) *Stereochemistry of Organic Electrode Processes* in *Organic Electrochemistry*, 4th edn (eds H. Lund and O. Hammerich), Marcel Dekker, New York, Chapter 26. (b) Fuchigami, T. and Inagi, S. (2014) *Stereochemistry of Organic Electrode Processes* in *Organic Electrochemistry*, 5th edn (eds O. Hammerich and B. Speiser), Taylor & Francis, Chapter 27.
34. Grimshaw, J. (2000) *Electrochemical Reactions Mechanisms in Organic Chemistry*, Elsevier, Amsterdam, pp. 80–83, 268–269.
35. Gaurley, R.N., Grimshaw, J. and Millar, P.G. (1967) *Chem. Commun.*, 1278–1279.
36. Kopilov, J., Kariv, E. and Miller, L.L. (1977) *J. Am. Chem. Soc.*, **99**, 3450–3454.
37. Watkins, B.F., Behling, J.R., Kariv, E. and Miller, L.L. (1975) *J. Am. Chem. Soc.*, **97**, 3549–3550.
38. Firth, B.E., Miller, L.L., Mitani, M., Rogers, T., Lennox, J. and Murray, R.W. (1976) *J. Am. Chem. Soc.*, **98**, 8271–8272.
39. Abe, S., Nonaka, T. and Fuchigami, T. (1983) *J. Am. Chem. Soc.*, **105**, 3630–3632.
40. Osa, T., Kashiwagi, Y., Yanagisawa, Y. and Bobbitt, J.M. (1994) *J. Chem. Soc., Chem. Commun.*, 2535–2537.
41. Osa, T. (1998), *Construction of New Mediator Systems* in *New Challenges* in *Organic Electrochemistry* (ed. T. Osa), Gordon & Breach, Amsterdam, pp. 183–219.
42. Steckhan, E. (1994) *Topics in Current Chemistry 170. Electrochemistry V*, Springer-Verlag, 84–111.
43. (a) Höllrigl, V., Otto, K. and Schmid, A. (2007) *Adv. Synth. Catal.*, **349**, 1337–1340. (b) Huang, J., Fu, X., Wang, G., Ge, Y. and Miao, Q. (2012) *J. Mol. Catalysis A: Chemical.*, **357**, 162–173.
44. (a) Kashiwagi, Y., Kurashima, F., Chiba, S., Anzai, J., Osa, T. and Bobbitt, T.M. (2003) *Chem. Commun.*, 114–115. (b) Shiigi, H., Tanaka, T., Demizu, Y. and Onomura, O. (2008) *Tetrahedron Lett.*, **49**, 5247–5251. (c) Demizu, Y., Shiigi, H., Mori, H., Matsumoto, K. and Onomura, O. (2008) *Tetrahedron Asymmetry*, **19**, 2659–2665. (d) Minato, D., Arimoto, H., Nagasue, Y., Demizu, Y. and Onomura, O. (2008) *Tetrahedron*, **64**, 6675–6683.
45. (a) Merz, A. (1989) *Chemically Modified Electrodes* in *Topics in Current Chemistry, 152. Electrochemistry IV* (ed. E. Steckhan), Springer-Verlag, Berlin, pp. 49–90. (b) Murray, R.W. (1980) *Acc. Chem. Res.*, **13**, 135–141. (c) Nonaka, T. and Fuchigami, T. (1985) *J. Org. Synth. Chem.*, **43**, 565–574.
46. Kunugi, Y., Fuchigami, T., Tien, H.-J. and Nonaka, T. (1989) *Chem. Lett.*, 753–756.
47. Facci, J.S., Falcino, P.A. and Gold, J.M. (1986) *Langmuir*, **2**, 732–738.
48. Kokkinidis, G. (1986) *J. Electroanal. Chem.*, **201**, 217–236.
49. Jannakoudakis, A.D. and Kokkinidis, G. (1982) *Electrochim. Acta*, **27**, 1199–1205.

50. (a) Banno, N., Nakanishi, T., Matsunaga, M., Asahi, T. and Osaka, T. (2004) *J. Am. Chem. Soc.*, **126**, 428–429. (b) Nakanishi, T., Matsunaga, M., Nagasaka, M., Asahi, T. and Osaka, T. (2006) *J. Am. Chem. Soc.*, **128**, 13322–13323.
51. (a) Pinson, J. and Podvorica, F. (2005) *Chem. Soc. Rev.*, 429–439. (b) Mayers, B.T. and Fry, A.J. (2006) *Org. Lett.*, **8**, 41–414.
52. Baizer, M.M. and Hallcher, R.C. (1976) *J. Electrochem. Soc.*, **123**, 809–813.
53. Torii, S., Tanaka, H. and Ukida, M. (1979) *J. Org. Chem.*, **44**, 1554–1557.
54. Ishifune, M., Yamashita, H., Matsuda, M., Ishida, H., Yamashita, N., Kera, Y., Kashimura, S., Masuda, H. and Murase, H. (2001) *Electrochim. Acta*, **46**, 3259–3264.
55. Uneyama, K. and Watanabe, S. (1990) *J. Org. Chem.*, **55**, 3909–3912.
56. Li, W., Nonaka, T. and Chou, T.-C. (1999) *Electrochemistry*, **67**, 4–10.
57. Hallcher, R.C., Goodin, R.D. and Baizer, M.M. (1981) US Patent 429 3393.
58. (a) Li, W. and Nonaka, T. (1999) *J. Electrochem. Soc.*, **146**, 592–599. (b) Shen, Y., Atobe, M., Li, W. and Nonaka, T. (2003) *Electrochim. Acta*, **48**, 1041–1046.
59. Inagi, S., Nagai, H., Tomita, I. and Fuchigami, T. (2013) *Angew. Chem. Int. Ed.*, **52**, 6616–6619.
60. Chaussard, J., Folest, J.C., Nedelec, J.Y., Perichon, J., Sibille, S. and Troupe, M. (1990) *Synthesis*, 369–381.
61. (a) Silvestri, G., Gambino, S., Filardo, G. and Gulotta, A. (1984) *Angew. Chem. Int. Ed. Engl.*, **23**, 979–980. (b) Yamauchi, Y., Hara, S. and Senboku, H. (2010) *Tetrahedron*, **66**, 473–479. (c) Ohkoshi, M., Michinishi, J., Hara, S. and Senboku, S.H. (2010) *Tetrahedron*, **66**, 7732–7737.
62. Shono, T., Kashimura, S., Ishifune, M. and Nishida, R. (1990) *J. Chem. Soc. Chem. Comm.*, 1160–1161.
63. (a) Lehmkuhl, H. (1973) *Synthesis*, 377–396. (b) Tuck, D.G. (1979) *Pure Appl. Chem.*, **51**, 2005–2018.
64. Sibilie, S., d'Incan, E., Leport, L. and Peichon, J. (1986) *Tetrahedron Lett.*, **27**, 3129–3132.
65. Yamamoto, Y., Goda, S., Maekawa, H. and Nishiguchi, I. (2003) *Org. Lett.*, **15**, 2755–2758.
66. (a) Simons, J.H. (1949) *J. Electrochem. Soc.*, **95**, 47–67. (b) Rudge, A.J. (1971), *Electrochemical Fluorination* in *Industrial Electrochemical Processes* (ed. A.T. Kuhn), Elsevier, London, pp. 71–88. (c) Tasaka, A. (2004), *Electrochemical Perfluorination* in *Current Topics in Electrochemistry*, Vol. 10, Springer-Verlag, Berlin, pp. 1–36.
67. Rozhkov, I.N. (1976) *Russ. Chem. Rev.*, **45**, 615–629.
68. Rozhkov, I.N. (1983) *Organic Electrochemistry*, 2nd edn (eds M.M. Baizer and H. Lund), Marcel Dekker, New York, Chapter 24.
69. Childs, W.V., Christensen, L., Klink, F.W. and Kolpin, C.F. (1991) *Anodic Fluorination* in *Organic Electrochemistry*, 3rd edn (eds H. Lund and M.M. Baizer), Marcel Dekker, New York, Chapter 26.
70. Fuchigami, T. (1999) Electrochemistry applied to the synthesis of fluorinated organic substances, in *Advances in Electron Transfer Chemistry*, Vol. 6 (ed. P.S. Mariano), JAI Press, Greenwich CT.
71. Fuchigami, T. (2000) Electrochemical partial fluorination, in *Organic Electrochemistry*, 4th edn (eds H. Lund and O. Hammerich), Marcel Dekker, New York.

72. Fuchigami, T. and Inagi, S. (2011) *Chem. Commun.*, **47**, 10211–10223.
73. (a) Andres, D.F., Laurent, E.G., Marquet, B.S., Benotmane, H. and Bensadat, A. (1995) *Tetrahedron*, **51**, 2605–2618. (b) Laurent, E.G., Tardivel, R., Benotmane, H. and Bensadat, A. (1990) *Bull. Soc. Chim. Fr.*, **127**, 468–475. (c) Dmowski, W. and Kozlowski, T. (1997) *Electrochim. Acta*, **42**, 513–523.
74. (a) Laurent, E.G., Tardivel, R. and Thiebault, H. (1983) *Tetrahedron Lett.*, **24**, 903. (b) Ventalon, F.M., Faure, R., Laurent, E.G. and Marquet, B.S. (1994) *Tetrahedron: Asymmetry*, **5**, 1909–1912.
75. Laurent, E.G., Marquet, B., Tardivel, R. and Thiebault, H. (1987) *Tetrahedron Lett.*, **28**, 2359–2362.
76. Fuchigami, T., Shimojo, M., Konno, A. and Nakagawa, K. (1990) *J. Org. Chem.*, **55**, 6074–6075.
77. Konno, A., Nakagawa, K. and Fuchigami, T. (1991) *J. Chem. Soc., Chem. Commun.*, 1027–1029.
78. Fuchigami, T., Konno, A., Nakagawa, K. and Shimojo, M. (1994) *J. Org. Chem.*, **59**, 5937–5941.
79. Dawood, M. (2004) *Tetrahedron*, **60**, 1435–1451.
80. Hou, Y., Higashiya, S. and Fuchigami, T. (1999) *J. Org. Chem.*, **64**, 3346–3349.
81. Shaaban, M.R., Inagi, S. and Fuchigami, T. (2009) *Electrochim. Acta*, **54**, 2635–2639.
82. Fuchigami, T., Narizuka, S. and Konno, A. (1992) *J. Org. Chem.*, **57**, 3755–3757.
83. Narizuka, S. and Fuchigami, T. (1993) *J. Org. Chem.*, **58**, 4200–4201.
84. Cao, Y., Hidaka, A., Tajima, T. and Fuchigami, T. (2005) *J. Org. Chem.*, **70**, 9614–9617.
85. Ishii, H., Yamada, N. and Fuchigami, T. (2000) *Chem. Commun.*, 1617–1618.
86. (a) Fuchigami, T. and Sano, M. (1996) *J. Electroanal. Chem.*, **414**, 81–84. (b) Fuchigami, T., Tetsu, M. and Tajima, T. (2001) *Synlett*, 1269–1271. (c) Fuchigami, T. and Fujita, T. (1994) *J. Org. Chem.*, **59**, 1269–1271. (d) Fujita, T. and Fuchigami, T. (1996) *Tetrahedron Lett.*, **37**, 4725–4728.
87. (a) Dawood, K.M., Ishii, H. and Fuchigami, T. (1999) *J. Org. Chem.*, **64**, 7935–7939. (b) Shaaban, M.R., Ishii, H. and Fuchigami, T. (2000) *J. Org. Chem.*, **65**, 8685–8689. (c) Hou, Y. and Fuchigami, T. (2000) *J. Electrochem. Soc.*, **147**, 4567–4572.
88. Sawamura, T., Takahashi, K., Inagi, S. and Fuchigami, T. (2012) *Angew. Chem. Int. Ed.*, **51**, 4413–4416.
89. (a) Skotheim, T.A. and Reynolds, J.R. (eds) (2007) *Handbook of Conducting Polymers*, 3rd edn, CRC Press, Boca Raton, FL. (b) Inzelt, G. (2008) *Conducting Polymers*, Springer, Heidelberg.
90. (a) Heinze, J., Frontana-Uribe, B.A. and Ludwigs, S. (2010) *Chem. Rev.*, **110**, 4724–4771. (b) Beaujuge, P.M. and Reynolds, J.R. (2010) *Chem. Rev.*, **110**, 268–320.
91. (a) Bhadani, S.N. and Parravano, G. (1983), *Electrochemical Polymerization* in *Organic Electrochemistry*, 2nd edn, Marcel Dekker, Chapter 13. (b) Funt, B.L. (1990), *Electrochemical Polymerization* in *Organic Electrochemistry*, 3rd edn, Marcel Dekker, New York, Chapter 32.
92. Ando, S. and Ueda, M. (2002) *Synth. Met.*, **129**, 207–213.
93. Lemaire, M., Büchner, W., Garreau, R., Hoa, K.A., Guyard, A. and Roncali, J. (1990) *J. Electroanal. Chem.*, **281**, 293–298.

94. Sekiguchi, K., Atobe, M. and Fuchigami, T. (2002) *Electrochem. Commun.*, **4**, 881–885.
95. Shi, G., Jin, S., Xue, G. and Li, C. (2005) *Prog. Polym. Sci.*, **30**, 783–811.
96. Saito, N., Kanbara, T., Nakamura, Y., Yamamoto, T. and Kubota, K. (1994) *Macromolecules*, **27**, 756–761.
97. Utley, J.H.P. and Gruber, J. (2002) *J. Mater. Chem.*, **12**, 1613–1624.
98. Miller, R.D. and Michl, J. (1989) *Chem. Rev.*, **89**, 1359–1410.
99. Shono, T., Kashimura, S., Ishifune, M. and Nishida, R. (1990) *J. Chem. Soc., Chem. Commun.*, 1160–1161.
100. (a) Kashimura, S., Ishifune, M., Yamashita, N., Bu, H.B., Takebayashi, M., Kitajima, S., Yoshizawa, D., Kataoka, Y., Nishida, R., Kawasaki, S., Murase, H. and Shono, T. (1999) *J. Org. Chem.*, **64**, 6615–6621. (b) Shono, T., Kashimura, S. and Murase, H. (1992) *J. Chem. Soc., Chem. Commun.*, 896–897.
101. (a) Matyjaszewski, K. and Xia, J. (2001) *Chem. Rev.*, **101**, 2921–2990. (b) Kamigaito, M., Ando, T. and Sawamoto, M. (2001) *Chem. Rev.*, **101**, 3689–3745.
102. Magenau, A.J.D., Strandwitz, N.C., Gennaro, A. and Matyjaszewski, K. (2011) *Science*, **332**, 81.

6

New Methodology of Organic Electrochemical Synthesis

Toshio Fuchigami, Mahito Atobe and Shinsuke Inagi

In the 21st century, environmentally friendly processes are becoming much more important. The concept and significance of green sustainable chemistry (GSC), which was developed in the USA, has been recognized throughout the world, and nowadays new processes cannot be developed without consideration of GSC. Since the latter stages of the last century, much attention has been paid to organic electrosynthesis as a typical environmentally friendly process. Furthermore, in this century, electrolytic reactions with much lower emission processes and electrochemical methodologies based on new concepts have been developed. These related studies have become an active research area [1].

In this chapter, new methodologies for organic electrolysis are described. Even though some methodologies are not very new, they are mentioned because they are highly useful from the viewpoint of GSC.

Fundamentals and Applications of Organic Electrochemistry: Synthesis, Materials, Devices, First Edition. Toshio Fuchigami, Mahito Atobe and Shinsuke Inagi.

6.1 SPE ELECTROLYSIS AND ITS APPLICATIONS

In contrast to organic synthesis, electrolytic reactions require supporting salts. It is troublesome and time-consuming to separate the desired product from an electrolytic solution containing a large amount of supporting salts. Moreover, supporting salts are usually not recovered and recycled, and become waste after electrolysis. In general, supporting salts are very soluble in polar solvents, but, depending on the starting organic substrates, some are not soluble in polar solvents, only in non-polar solvents. As mentioned previously, the correct choice of supporting salt and electrolytic solvent is not easy. In sharp contrast, electrolysis using solid polymer electrolytes (SPE) does not require an electrolytic solution, therefore these problems are readily solved [2,3].

6.1.1 Principle of SPE Electrolysis

By chemical plating, fine metal particles are deposited on the surface of ion-exchange membranes like Nafion®, and then a wire mesh electrode or porous electrode made of the same metal is pressed on the opposite surface of the membrane to provide an SPE composite electrode, which has dual roles of electrode and supporting salt.

The general advantages of the SPE electrolysis system for organic synthesis can be described as follows: (i) to economize the separation and recycling of a supporting salt and (ii) to avoid any contamination or side reaction with a supporting electrolyte. There are two types of SPE systems. In one an electrode is pressed onto only one side of the surface of the ion-exchange membrane and in the other two electrodes are pressed on both surfaces. Figure 6.1 illustrates the latter SPE system. Substrate AH_2 is oxidized at the anode and the resulting protons migrate through the cation exchange membrane to the cathode. Thus, ionic conductivity, namely proton conductivity, is available. The arriving protons are reduced at the cathode to generate hydrogen gas solely in the absence of substrate (Figure 6.1a), while substrate B is reduced to BH_2 in the presence of substrate B (Figure 6.1b). In the latter case, two products, A and BH_2, are obtained from two starting substrates, AH_2 and B, respectively, therefore this electrolysis can be considered to be a kind of paired electrosynthesis.

In the case of the SPE electrolysis system in which an electrode is pressed onto only a single side of the surface of an ion-exchange membrane, the counter-electrode chamber is filled with electrolytic solvent. In SPE systems,

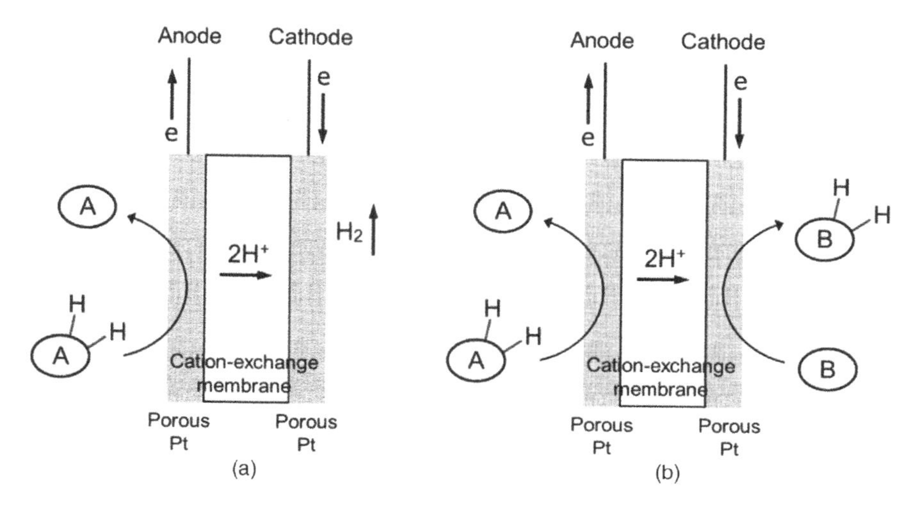

Figure 6.1 Principle of electrolysis using an SPE system with platinum electrodes

even non-polar solvents like hexane can be used as electrolytic solvents. Furthermore, when the starting substrate is a liquid or gas, electrolysis can be performed without any solvents. Moreover, the amount of redox mediator used can be reduced by immobilizing it on the SPE composite electrode.

Various organic electrosyntheses using SPE systems have been reported to date, e.g. hydrogenation of olefins, reduction of nitrobenzene, Kolbe electrolysis, dimethoxylation of furan and ketone synthesis using a halogen mediator, etc. [2–4]. SPE electrosynthesis is an energy-efficient electrolytic method since the cell voltage is low owing to the small distance between electrodes. However, the durability of SPE composite electrodes is not high and contamination of the SPE membrane decreases the current efficiency and yield. Those problems have not yet been solved and a commercialized SPE organic electrosynthesis has not been established to date. The original concept for SPE organic electrosynthesis is a polymer electrolyte fuel cell (PEFC). Thus, the development of new materials such as membranes and increased knowledge of fuel cell research should stimulate the progress of SPE electrosynthesis.

6.1.2 SPE Electrolysis with Cogeneration (Chemicals Production Using Fuel Cell Reactions)

The SPE electrolysis system has been applied to commercialized water electrolysis (water splitting) and fuel cells. A fuel cell is a device to convert

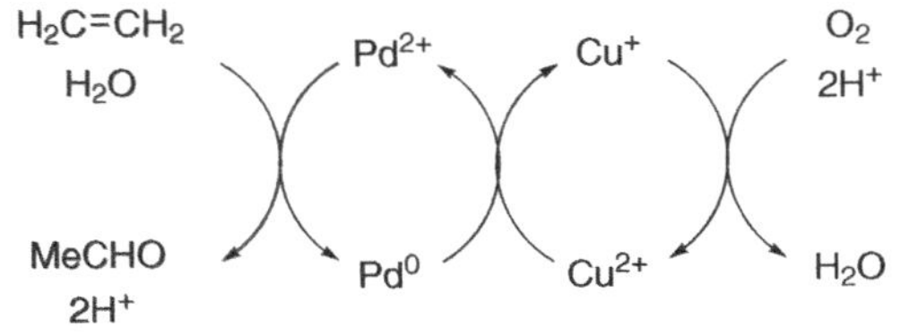

H2C=CH2
H2O
Pd2+
Cu+
O2
2H+
MeCHO
2H+
Pd0
Cu2+
H2O

Figure 6.2 Reaction mechanism of the Wacker oxidation reaction

Gibbs free energy in a chemical reaction into electricity through electrochemical reactions. In an H_2–O_2 fuel cell, electricity is obtained through the formation of water from oxygen and hydrogen. This principle suggests that catalytic oxidation and reduction in chemical synthesis can be converted to fuel cell reactions at an anode and cathode. For instance, the Wacker oxidation of ethylene is catalyzed by redox species like Pd^{2+}/Pd^0 and Cu^{2+}/Cu^+, as shown in Figure 6.2. Based on the flows of protons and electrons in the reaction scheme, it can be expected that the Wacker oxidation of ethylene to acetaldehyde with oxygen would be possible using fuel cell reactions, as shown in Figure 6.3.

In fact, ethylene together with water vapour and oxygen are carried into the left and right compartments, respectively, as shown in Figure 6.3, and then the circuit is closed. Under these conditions, current flows and acetaldehyde is formed highly selectively (95%) at 100°C. Thus, in this system, both flow of electric current and autoxidation of ethylene take place spontaneously even if potential is not applied. This fuel cell system enables cogeneration of electricity and chemicals [5].

Anode
Cathode
e
e
MeCHO
H2O
CH2=CH2, H2O
H+
O2
Cation-exchange
membrane
Porous
Pt
Porous
Pt

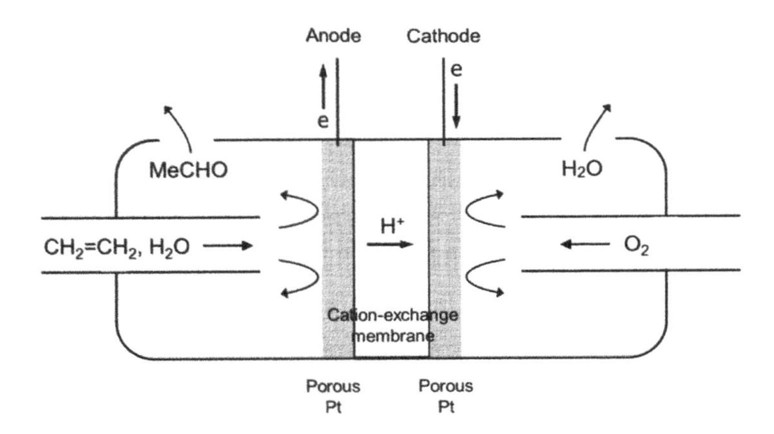

Figure 6.3 Fuel cell reaction

6.2 ELECTROLYTIC SYSTEMS USING SOLID BASES AND ACIDS

Ordinary soluble bases are readily oxidized anodically, but insoluble solid-supported bases are not oxidized at the anode. Based on this fact, new electrolytic systems have recently been developed using recyclable and reusable solid-supported bases and acids as supporting electrolytes [5–8]. For instance, as shown in Figure 6.4, solid (silica gel or porous polystyrene)-supported bases like piperidine dissociate protic organic solvents like methanol and acetic acid to protons and anions, and the resulting protons act as the main carriers of electronic charge.

This system is widely applicable to various anodic nucleophilic substitutions, such as methoxylation and acetoxylation of various compounds (Eq. 6.1). Moreover, it is notable that after the electrolysis, solid-supported bases are easily separated by only filtration and the product is readily isolated from the filtrate. The recovered solid-supported bases can be reused many times, as shown in Figure 6.5.

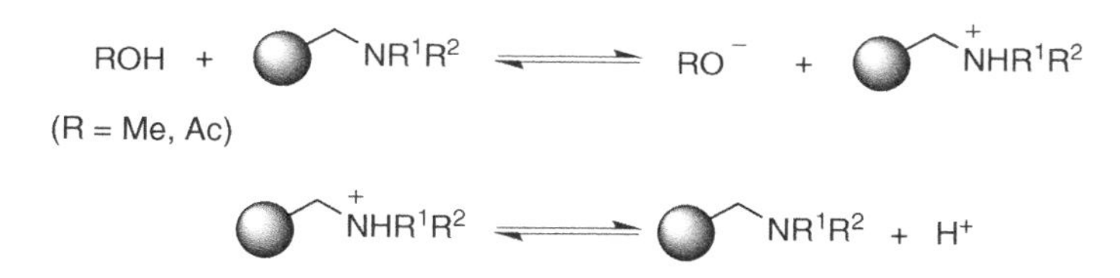

Figure 6.4 Dissociation of protic solvents by solid base

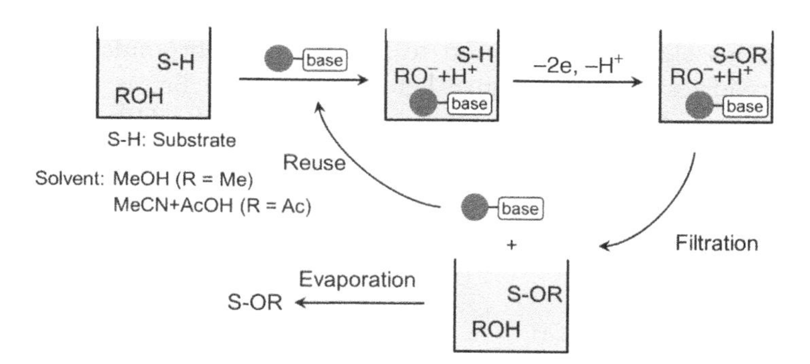

Figure 6.5 Recovery and recycle system of anodic methoxylation and acetoxylation using solid base

Substrate —(−2e, −H^+; Si-supported piperidine, MeOH)→ Product

PhS–CH$_2$–CF$_3$ → PhS–CH(OMe)–CF$_3$ (76%)

N-Boc pyrrolidine → 2-OMe N-Boc pyrrolidine (94%)

HO–C$_6$H$_4$–OMe → 4,4-dimethoxycyclohexa-2,5-dienone (94%)

(6.1)

6.3 SOLID-SUPPORTED MEDIATORS

Although mediators have many advantages as explained in Chapter 2, 3, and 5, they have to be removed from the electrolytic solution after electrolysis. However, mediators have never been recovered after electrolysis for a long time. Quite recently, disposal-type mediators have become a problem from GSC, particularly atom economical aspect. In order to overcome such problems, easily separable and recyclable mediators have been developed.

At an early stage of this approach, a cross-linked poly-4-vinylpyridine hydrobromide as a solid-supported mediatory system was developed [9]. As shown Eq. 6.2, Br^- derived from the solid base is anodically oxidized in an aqueous electrolytic solution to generate hypobromide ion, BrO^-, which oxidizes alcohols to give ketones, and then Br^- is regenerated. Thus, the mediator can be easily recovered as the HBr salt of the solid base.

$$\text{Py}^+\text{NHBr}^- + H_2O \xrightarrow{-2e} \text{Py}^+\text{NHOBr}^- + H_2$$

$$\text{Py}^+\text{NHOBr}^- + RCH(OH)R' \longrightarrow \text{Py}^+\text{NHBr}^- + RC(=O)R' + H_2O$$

(6.2)

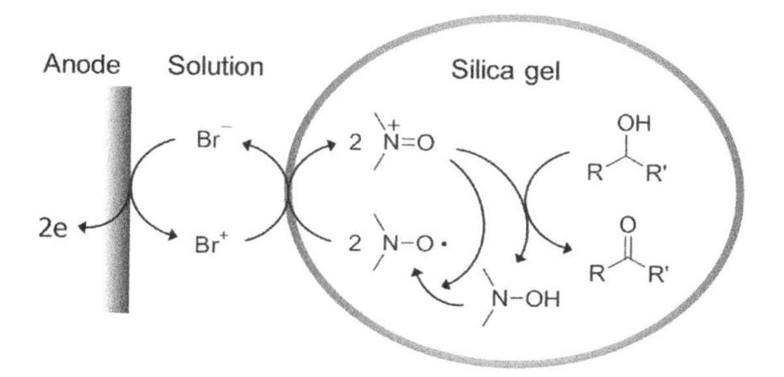

Figure 6.6 Indirect oxidation of alcohols with *N*-oxyl-immobilized silica gel

Since *N*-oxyl radicals as represented by TEMPO generally show reversible redox properties, as shown in Eq. 6.3, they are useful mediators for the oxidation of alcohols. *N*-Oxyl adsorbed on the surface of silica gel disperse mediatory system has been developed (Figure 6.6) [10].

$$\text{N–OH} \underset{e,\,H^+}{\overset{-e,\,-H^+}{\rightleftharpoons}} \text{N–O}^\bullet \underset{e}{\overset{-e}{\rightleftharpoons}} \text{N}^+\text{=O} \qquad (6.3)$$

In this dispersive system, Br^- is at first anodically oxidized to generate Br^+, which oxidizes *N*-oxyl on the silica gel to form oxoammonium ion ($N{=}O^+$) followed by oxidation of alcohols to produce ketones. In this case, the oxidation can be achieved without organic solvent. One of advantages of the dispersed system is as follows. After electrolysis, electrolytic solution is filtered and the remaining silica gel is washed with appropriate solvents like acetone to provide the product. However, during washing, the mediator also dissolves into the organic solvent. In order to avoid dissolution of the mediator, *N*-oxyl-immobilized silica gel was developed as a readily recyclable mediator [11]. *N*-Oxyl-immobilized polymer particles have also been developed. Furthermore, the combination of electrodes modified with TEMPO-immobilized polymer and chiral bases is known to be highly useful for asymmetric synthesis [12] (see also section 5.3).

6.4 BIPHASIC ELECTROLYTIC SYSTEMS

In general, electrochemical reactions are carried out in homogeneous electrolytic solutions. However, the electrochemical reaction can be conducted even in heterogeneous electrolytic solutions unless both electron transfer and mass transfer processes are not inhibited by the reaction system. In this section, we will discuss biphasic electrolytic systems such as emulsion electrolysis, suspension electrolysis, electrolysis using phase transfer catalysis and the thermomorphilic biphase electrolytic reaction system [13–15].

6.4.1 Emulsion Electrolysis

As shown in Figure 6.7, when the product is an oxidizable and reducible compound such as an aldehyde, an organic phase in a biphasic electrolytic system (in general, a system composed of organic and water phases in which organic droplets are finely dispersed into water (aqueous electrolyte)) is often used to extract the product in order to prevent its over-oxidation or over-reduction at the electrode.

On the other hand, immiscible organic compounds are often electrolyzed in their emulsion. However, in these cases, the emulsion contributes to maintain the saturated concentration of immiscible organic substrate in an aqueous electrolyte and the emulsified droplet itself is not directly electrolyzed at the electrode.

Unlike the above emulsion electrolytic systems, the direct electrolysis of the emulsified droplets can also take place at the electrode if the droplets contain supporting electrolytes and their presence allows the formation of an electric bilayer inside the droplets. However, in this case, droplets should be miniaturized to the sub-micrometre range by the use of special surfactants or ultrasonication in order to obtain a practical reaction rate [13].

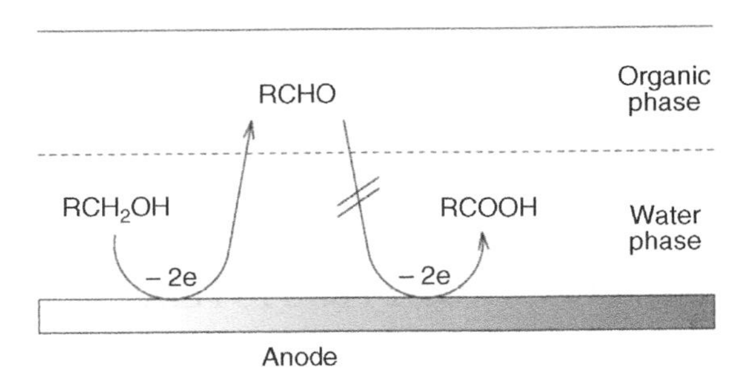

Figure 6.7 Emulsion electrolysis

6.4.2 Suspension Electrolysis

The direct electrolysis of suspensions of non-electronconductive solid materials such as most organic compounds cannot take place in principle, since electrochemical electron transfer between a solid electrode and a non-conductive solid particle hardly ever occurs even when they contact with each other. On the other hand, the indirect electrolysis of suspended solid particles using soluble redox mediators may be able to take place, since this consists of a series sequence of two possible electrochemical and chemical electron-transfer steps between solid and liquid phases (electrode (solid)–dissolved mediator (liquid)–suspending particle (solid)) (Figure 6.8) [14]. Similarly to this case, the electrolysis of emulsified liquid droplets can be conducted by using electron-transfer mediators.

6.4.3 Electrolysis Using Phase-Transfer Catalysis

Anodic aromatic substitutions have been carried out using dispersions of an organic solvent (usually methylene chloride) in aqueous media. Current flow takes place mainly through the relatively conducting aqueous phase (thereby lowering power costs) while the synthetic reactions have been assumed to be confined to the organic phase; the function of the phase-transfer catalyst is to transfer with anions into the organic phase both to confer adequate conductivity to this phase and to provide the coupling agent (e.g. CN^- or CH_3COO^-) for the intermediates generated from the aromatic substrates (Figure 6.9).

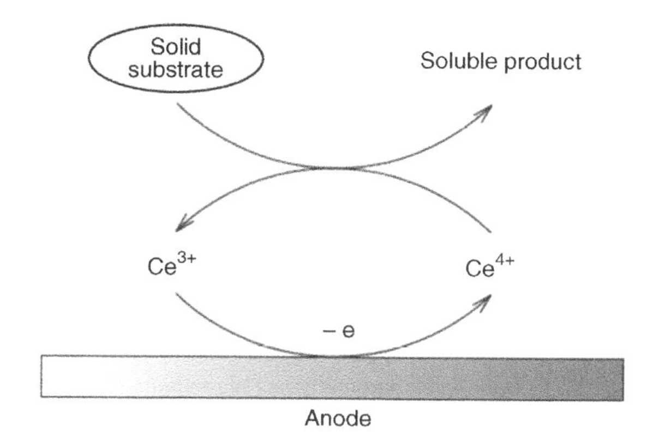

Figure 6.8 Suspension electrolysis

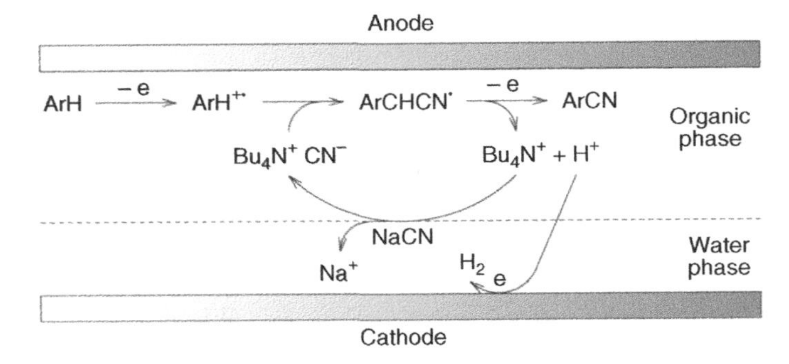

Figure 6.9 Example of electrolysis using phase-transfer catalysis

6.4.4 Thermomorphic Biphasic Electrochemical Reaction System

Chiba and co-workers constructed thermomorphic biphasic electrochemical reaction systems composed of organic solvents of different polarity (e.g. cyclohexane–nitroalkanes) [15]. As shown in Figure 6.10, in the

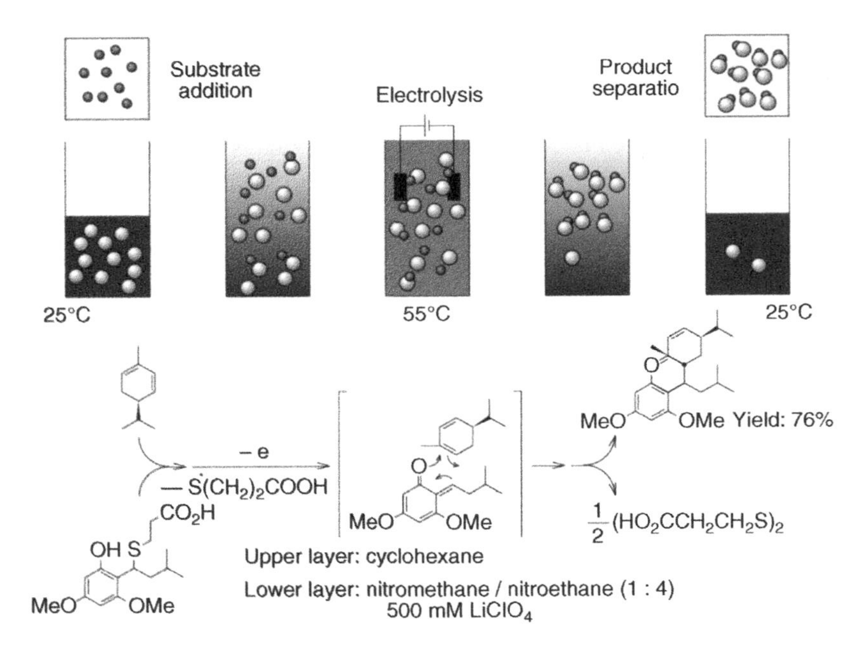

Figure 6.10 Thermomorphic biphasic electrochemical reaction system

thermally immixed homogeneous electrolyte solution, the anodic desulfurization proceeds smoothly to give *o*-quinone methide, and then reacts with terpene to give the cycloadduct as a final product. After the completion of the electrolysis, the product as the cyclohexane solution can be separated from the polar electrolytic solution (nitroalkane phase) simply by on-cooling phase separation. In addition, the separated polar electrolytic solution can be reused for the next cycle of the reaction.

6.5 CATION POOL METHOD

Electro-oxidatively generated carbocations can react with nucleophiles to form covalent bonds. Because of the short life-time of carbocations, it is necessary to generate carbocations in the presence of nucleophiles, followed by immediate reaction. In the above system, the nucleophiles used should have higher oxidation potential than precursor molecules of carbocation. Since the carbon nucleophile is easily oxidized, the formation of carbon–carbon bonds using carbocations generated by anodic oxidation is generally difficult.

The cation pool method can preserve carbocations stabilized by neighbouring heteroatoms or aromatics at low temperature (–78°C) without decomposition. Addition of nucleophiles to the cation pool results in desirable nucleophilic reaction (Figure 6.11) [16,17].

At low temperature, the viscosity of the electrolytic solution increases and this decreases the ionic conductivity of the solution. Dichloromethane is a suitable solvent for the cation pool method that has sufficient conductivity for electrolysis. With a divided electrolytic cell, the substrate is injected into the anode chamber and trifluoromethane sulfonic acid (the proton source) is injected into the cathodic chamber to promote a sacrificial cathodic reaction.

A variety of cation pools are successfully generated, as shown in Figure 6.12. The cation pool of alkoxycarbenium ions is obtained by

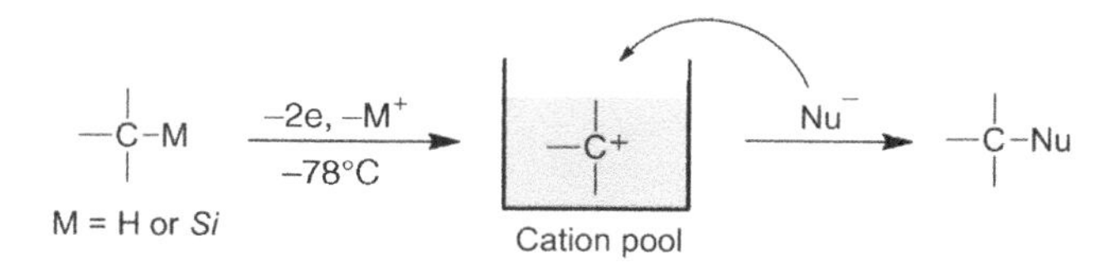

Figure 6.11 Nucleophilic substitution reaction by the cation pool method

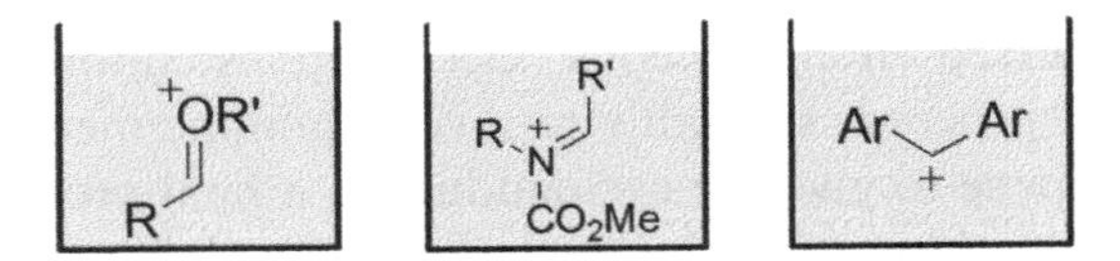

Figure 6.12 Examples of cation pools

anodic oxidation of α-silyl ethers at low temperature, and the addition of allylsilanes as carbon nucleophiles yields the desired products efficiently. Similarly, an *N*-acyliminium ion pool and a diarylcarbenium ion pool are available for the resulting chemical transformations. Figure 6.13 shows examples of nucleophilic reactions of accumulated *N*-acylcarbenium ion species.

The concept of the cation pool method can be extended to a cation flow method in combination with a micro-reactor system (see section 6.8). This method makes it possible not only to survey short-life intermediates generated electrochemically but also to realize the combinatorial synthesis of useful materials.

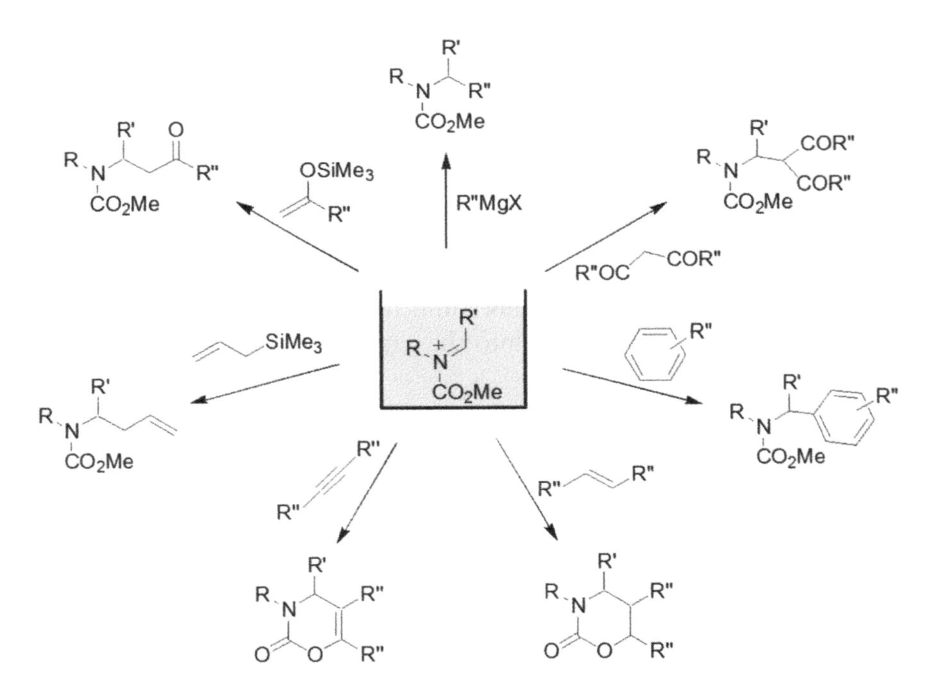

Figure 6.13 Various reactions of the *N*-acyliminium cation ion pool

6.6 TEMPLATE-DIRECTED METHODS

In general, a regioselective intermolecular coupling reaction is difficult compared to an intramolecular one. Template-directed reactions via intramolecular reactions enable selective coupling reactions that are otherwise difficult to achieve. As shown in Figure 6.4, regioselective anodic coupling of phenols can be achieved using a boron template [18]. First, the template, tetraphenoxyborate, is prepared in a one-pot-procedure and then its anodic oxidation followed by hydrolysis provides an ortho-ortho coupled product selectively.

OH — Na, $B(OH)_3$ → [tetraaryloxyborate B^- Na^+] — Anodic oxidation, MeCN → HO / OH biaryl, 85%

(6.4)

A silicon template is also known [19].

6.7 ELECTROLYSIS IN SUPERCRITICAL FLUIDS

The commercialization of electrosynthetic processes has been restricted by the limited solubility of substrates and products in conventional electrolytic solutions, the poor interphase mass transport characteristics associated with the two-phase system in which the reaction occurs at solid (electrode)–liquid (electrolyte) interfaces, the low selectivity of the desired reaction products and the complex processing schemes often used to recover products.

Supercritical fluid solvents can overcome many of the limitations associated with conventional solvents such as water and organic solvents. Supercritical fluids have traditionally been applied to chromatography, extraction, cleaning and so on [20]. Moreover, they are becoming widely recognized as useful media for organic and polymer syntheses in a range of laboratory and industrial processes because of their low toxicity, ease of solvent removal, potential for recycling and variation of reaction rates [21,22].

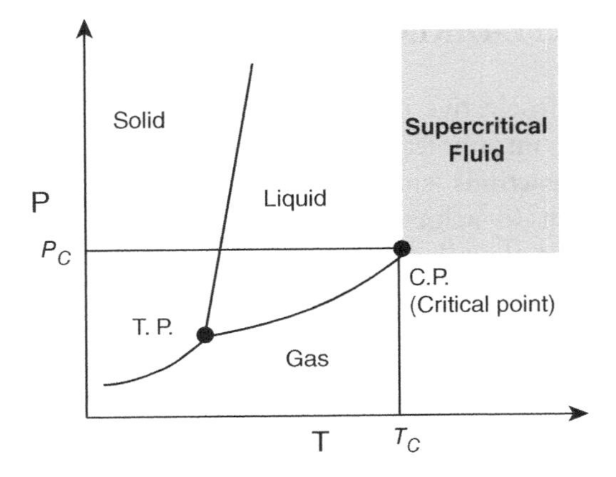

Figure 6.14 Pressure–temperature phase diagram

A supercritical fluid can be defined as any substance that is above its critical temperature (T_c) and critical pressure (P_c), and exists as a single phase (Figure 6.14). The physico-chemical properties of a supercritical fluid are between those of liquids and gases, and they can vary between those exhibited by gases to liquid-like values by small changes in pressure and temperature [20]. For example, supercritical fluids exhibit larger diffusivility and lower viscosity than conventional liquids. Consequently, interphase mass transfer resistance is also lower relative to the liquid solvent. On the other hand, substrates and products that are sparingly soluble in the liquid phase can become significantly more soluble in supercritical fluids. Such unique and useful properties of the fluids make them viable solvent media for electrosynthesis.

Conventional supercritical solvents, such as carbon dioxide and ethane, have mild critical conditions, but they lack the ability to solvate polar electrolytes. A reaction medium should therefore generally be designed by mixing with co-solvent such as methanol [23], ethanol [24], acetone [24], acetonitrile [25] or DMF [26] to produce a suitable electrolytic solution.

Dombro *et al.* reported that dimethyl carbonate has been electrosynthesized from carbon monoxide and methanol in a supercritical carbon dioxide–methanol medium (Eq. 6.5) [27]. They added 0.35 mole fraction of methanol to supercritical carbon dioxide to produce a mixture capable of dissolving an ammonium salt. Such a mixture is capable of ionic conductivity and the electrolysis proceeded smoothly to afford dimethyl carbonate in an excellent current efficiency (near 100%).

$$2CH_3OH + CO \xrightarrow[\substack{Bu_4NBF_4 \\ scCO_2\text{-}CH_3OH}]{\text{electrolysis, 90°C}} (CH_3O)_2CO \quad (6.5)$$

In another example, Tokuda *et al.* reported electrochemical carboxylation of benzyl chloride, cinnamyl chloride and 2-chloronaphtalene in supercritical carbon dioxide–DMF media (Eqs. 6.6–6.8) [26]. A mixture used in this work also exhibited the ability to solvate polar electrolytes like ammonium salts. In these reactions, since carbon dioxide plays the role of a reactant, CO_2-rich conditions such as supercritical conditions are suitable for improving product yields.

Benzyl chloride + $scCO_2$ (40°C, 80 kg cm^{-2}) → [(−) Pt | Mg (+), 3 F mol^{-1}, 37.5 mA cm^{-2}, Bu_4NBr / $scCO_2$ + DMF] → phenylacetic acid (COOH); Yield: 81 % (6.6)

Cinnamyl chloride + $scCO_2$ (46°C, 80 kg cm^{-2}) → [(−) Pt | Mg (+), 3 F mol^{-1}, 37.5 mA cm^{-2}, Bu_4NBr / $scCO_2$ + DMF] → product (COOH); Yield: 64 % (6.7)

2-(Chloromethyl)naphthalene + $scCO_2$ (43°C, 80 kg cm^{-2}) → [(−) Pt | Mg (+), 3 F mol^{-1}, 37.5 mA cm^{-2}, Bu_4NBr / $scCO_2$ + DMF] → product (COOH); Yield: 61 % (6.8)

As mentioned above, conventional supercritical fluids like supercritical carbon dioxide are the low dielectric constant of the fluid. Hence, many supporting electrolytes are usually hard to dissolve in the fluid. A polar organic solvent should therefore be added to the fluid to produce a suitable electrolytic solution. However, the inclusion of a co-solvent limits the accessible potential range and introduces complications. The use of

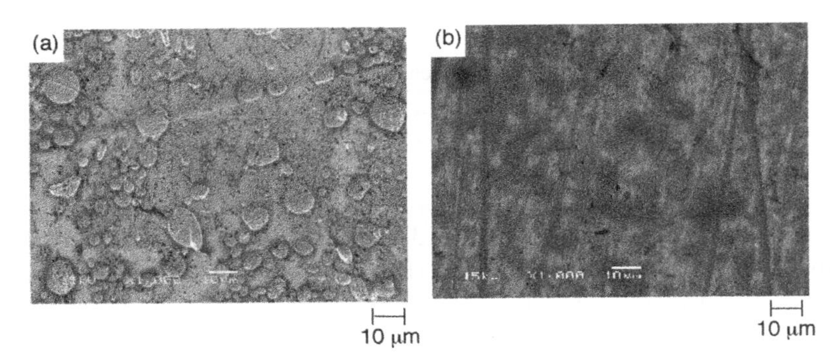

Figure 6.15 SEM photographs of polypyrrole films polymerized in (a) acetonitrile and (b) supercritical fluoroform solutions

supercritical fluorinated hydrocarbons is an attractive alternative which avoids these problems. For example, supercritical fluoroform exhibits relatively high solubility and its dielectric constant can be controlled from 1 to 7 by manipulating either the temperature or the pressure of $scCHF_3$ without any additives like polar solvents [28]. Supercritical fluoroform has therefore been used as the reaction medium in electrochemical syntheses.

Atobe *et al.* reported the electrochemical synthesis of polypyrrole and polythiophene in supercritical fluoroform [29]. Both the corresponding monomers (pyrrole and thiophene) can be electropolymerized faster in supercritical fluoroform than in a conventional organic media like acetonitrile solution, and the films obtained have a highly uniform structure (Figure 6.15).

Moreover, Atobe *et al.* successfully prepared a polythiophene nanobrush using template electrochemical polymerization in supercritical fluoroform (Figure 6.16) [30]. In this work, nanoporous alumina membranes

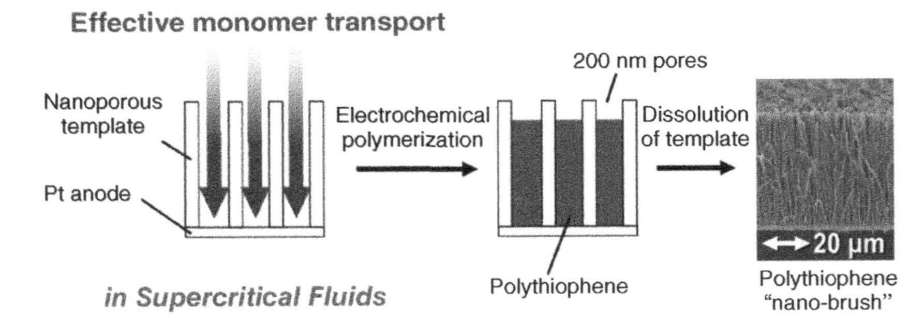

Figure 6.16 Preparation of polythiophenenano-brush using template electrochemical polymerization in supercritical fluoroform

(60 μm thick, 200 nm pore size) coated on one side with evaporated Pt (about 500 nm thick) were employed as a template electrode for polythiophene electrodeposition into pores. The use of the special properties of the fluids such as higher diffusivility and lower viscosity enabled effective monomer transport into the porous template and nanoprecise filling with the polymers. Consequently, the solid polythiophene nano-brush was obtained after removal of the alumina membrane.

6.8 ELECTROLYSIS IN IONIC LIQUIDS

Ionic liquids consist of cations and anions without any solvent, and they are in a liquid state around room temperature. Since ionic liquids have unique properties like non-flammability, thermal stability, non-volatility and reusability, they have been intensively studied as a green solvent from basic and applied aspects [31–35]. Since they have also good electro-conductivity, much attention has been paid to their application in organic electrochemical devices. In fact, intensive studies on applications to rechargeable lithium batteries, electric double-layer capacitors, wet solar cells and fuel cells have recently been carried out.

In this section, the structures and physical properties of ionic liquids are explained and their application to electrolytic reactions is described.

6.8.1 Structures of Ionic Liquids

Ionic liquids are roughly classified into two families based on kinds of cations. One consists of nitrogen-containing heteroaromatic cations and the other consists of aliphatic ammonium cations. Aliphatic ionic liquids are subdivided into open-chain and cyclic. Based on kinds of anions, ionic liquids are divided into five classes: chloroalminate, fluoroinorganic, fluoroorganic, poly(hydrogen fluoride) and non-fluoroionic. In addition, recently phosphine-containing ionic liquids like phosphonium salts have been prepared, and various novel types of ionic liquids with other new anions have been continuously developed (Figure 6.17).

6.8.2 Hydrophilicity and Hydrophobicity of Ionic Liquids

Ionic liquids can also be classified as hydrophilic or hydrophobic, as shown in Table 6.1. Regardless of cation type, ionic liquids having BF_4^-

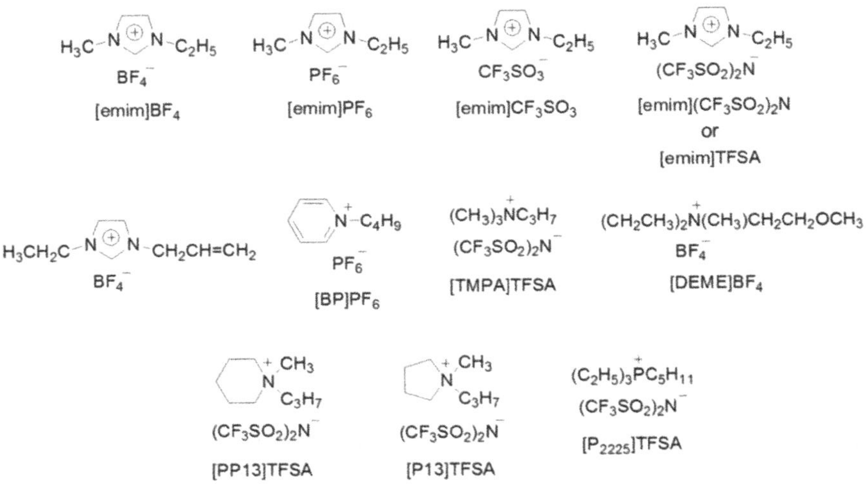

$H_3C-N^{\oplus}N-C_2H_5$ BF_4^- [emim]BF_4

$H_3C-N^{\oplus}N-C_2H_5$ PF_6^- [emim]PF_6

$H_3C-N^{\oplus}N-C_2H_5$ $CF_3SO_3^-$ [emim]CF_3SO_3

$H_3C-N^{\oplus}N-C_2H_5$ $(CF_3SO_2)_2N^-$ [emim]$(CF_3SO_2)_2N$ or [emim]TFSA

$H_3CH_2C-N^{\oplus}N-CH_2CH{=}CH_2$ BF_4^-

$\overset{+}{N}-C_4H_9$ PF_6^- [BP]PF_6

$(CH_3)_3\overset{+}{N}C_3H_7$ $(CF_3SO_2)_2N^-$ [TMPA]TFSA

$(CH_2CH_3)_2\overset{+}{N}(CH_3)CH_2CH_2OCH_3$ BF_4^- [DEME]BF_4

$\overset{+}{N}(CH_3)(C_3H_7)$ $(CF_3SO_2)_2N^-$ [PP13]TFSA

$\overset{+}{N}(CH_3)(C_3H_7)$ $(CF_3SO_2)_2N^-$ [P13]TFSA

$(C_2H_5)_3\overset{+}{P}C_5H_{11}$ $(CF_3SO_2)_2N^-$ [P_{2225}]TFSA

Figure 6.17 Typical examples of the structures of ionic liquids and their abbreviations

and $CF_3SO_3^-$ as an anion are hydrophilic while those with PF_6^- and $(CF_3SO_2)_2N^-$ are hydrophobic. As a characteristic property, hydrophobic ionic liquids are miscible with neither water nor ordinary organic solvents like ether and hexane, and hence they make phase separation resulting in formation of three phases. The products can therefore be separated by liquid–liquid extraction, which is one of the big advantages of ionic liquids.

In the case of the widely used imidazolium ionic liquids, the relationship between their molecular structures and physical properties, such as melting point, viscosity and electroconductivity, is as follows.

Table 6.1 Hydrophobic and hydrophilic properties of imidazolinium salt ionic liquids

Ionic liquid	Property
[emim]BF_4	Hydrophilic
[bmim]PF_6	Hydrophobic
[bmim]BF_4	Hydrophilic
[emim]$(CF_3SO_2)_2N$	Hydrophobic
[bmim]$(CF_3SO_2)_2N$	Hydrophobic
[bmim]CF_3SO_3	Hydrophilic
[emim]CH_3OSO_3	Hydrophilic
[emim]CF_3SO_3	Hydrophilic
[emim]NO_3	Hydrophilic

Me–N⊕N–Et $[BF_4]^-$　　　Me–N⊕N–C_6H_{13} $[BF_4]^-$

$[emim]BF_4$

Viscosity: 31.8 cP(25°C)
Melting point: 14.6°C
Conductivity: 13.6 mS/cm

Viscosity: 223.8 cP(25°C)
Melting point: –82°C
Conductivity: 1.04 mS/cm

Figure 6.18 Relationship between alkyl chain length of imidazolinium salt ionic liquids and physical properties (viscosity, melting point and electroconductivity)

1. With increasing alkyl chain length the viscosity increases while the melting point and electroconductivity decrease. However, the relationship between melting point and alkyl chain length is not always observed, as shown in Figure 6.18. In order to decrease the melting point an unsymmetrical or bulky cation should be used, which results in a weak columbic interaction between cation and anion.
2. When a methyl group is introduced to the 2-position of the imidazolium ring, hydrogen bonding at the 2-position cannot occur. The melting point does not decrease unexpectedly but increases. Thus, the relationship between hydrogen bonding and melting point has not yet been clarified.
3. When an alkyl group is introduced to the 2-position of the imidazolium ring, the reduction of the ring becomes difficult but oxidation becomes easier. Moreover, when the imidazolium ion as cation is changed to an aliphatic ammonium ion, the oxidative resistance increases, and the viscosity and melting point also increase. Regarding the viscosity, imidazolium ionic liquids have the lowest viscosity among various types of ionic liquids at the present time (Figure 6.19).

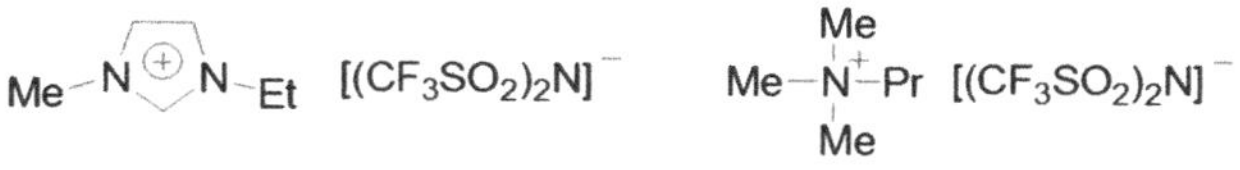

Me–N⊕N–Et $[(CF_3SO_2)_2N]^-$　　　Me–N^+(Me)(Me)–Pr $[(CF_3SO_2)_2N]^-$

$[emim](CF_3SO_2)_2N$

Viscosity: 28 cP(25°C)
Melting point: –16°C
Conductivity: 8.4 mS/cm

Viscosity: 72 cP(25°C)
Melting point: 19°C
Conductivity: 3.2 mS/cm

Figure 6.19 Relationship between cationic moieties of ionic liquids and physical properties (viscosity, melting point and electroconductivity)

6.8.3 Polarity of Ionic Liquids

The polarity of ionic liquids can be estimated by the energy difference, E_T^N between the ground state and the excited state of Reichrdt dye, which is commonly used for the polarity evaluation of ordinary solvents. The polarity of imidazolium ionic liquids and pyridinium ionic liquids is similar to that of ethanol, and is a little lower than that of methanol. However, their polarity is much higher than that of MeCN and DMF, therefore it is equivalent to that of protic solvents. The polarity of ionic liquids and organic solvents can be illustrated as follows:

$$\text{MeOH} > \text{imidazolium salt ionic liquid} > \text{pyridinium salt ionic liquid} = \text{EtOH} > \text{MeCN} > \text{DMF}$$

6.8.4 Electrochemical Properties of Ionic Liquids

Although the electroconductivity of a non-aqueous electrolyte is lower than that of an aqueous one, as shown in Table 6.2, non-aqueous electrolytes have the advantage of a wide usable potential window. Accordingly, ionic liquids belong to non-aqueous electrolytes.

As shown in Table 6.3, imidazolium ionic liquids generally have relatively good electroconductivity, while poly(hydrogen fluoride) salt ionic liquids exhibit rather good electroconductivity regardless of cation (Table 6.4). 1-Ethyl-3-methylimidazolium (emim) salts show the highest electroconductivity at room temperature among the series of ionic liquids. Furthermore, since the electroconductivity of ionic liquids is correlated with the mobility of ions, it greatly depends on the viscosity of ionic liquids. Accordingly, the molecular design of ionic liquids with low viscosity is of great importance. Aliphatic ammonium ionic liquids exhibit

Table 6.2 Classification of ionic liquids and electroconductivity

	Electrolyte		Electroconductivity (at 25°C) (mS cm^{-1})
Aqueous system	Acid	35 wt% H_2SO_4/H_2O	848
	Alkaline	30 wt% KOH/H_2O	625
	Neutral	30 wt% $ZnCl_2/H_2O$	105
Non-aqueous system	Organic	1 M $LiPF_6$/EC+EMC	9.6
	Inorganic	2 M $LiAlCl_4/SOCl_2$	20.5
	Ionic liquid	[emim]BF_4	13.6

Table 6.3 Physical and electrochemical properties of imidazolinium salt ionic liquids at 25°C

Ionic liquid	Melting point (°C)	Density (g cm^{-1})	Viscosity (mPa)	Conductivity (mS cm^{-1})	Potential window	
					E_{red}	E_{ox} (V vs. Li$^+$/Li)[c]
[emim]$AlCl_4$	8	1.29	18	22.6	1.0	5.5
[emim]$H_{2.3}F_{3.3}$	−90	1.14	5	100	1.5	5.3
[emim]BF_4	11	1.24	43	13	1.0	5.5
[emim]CF_3CO_2	−14	1.29[b]	35[a]	9.6[a]	1.0[d]	4.6[d]
[emim]CH_3SO_3	39	1.25	160	2.7	1.3[d]	4.9[d]
[emim]CF_3SO_3	−10	1.38	43	9.3	1.0	5.3
[emim]$(CF_3SO_2)_2N$	−15	1.52[b]	28	8.4	1.0	5.7
[emim]$(C_2F_5SO_2)_2N$	−1		61	3.4	0.9	5.8
[emim]$(CF_3SO_2)_3C$	39		181	1.7	1.0	6.0

[a] 20°C.
[b] 22°C.
[c] GC (glassy carbon), 1 mA cm^{-2}, 20 mV s^{-1}.
[d] Pt, 50 mV s^{-1}.

lower conductivity since they generally have a higher viscosity compared to imidazolium ionic liquids. In contrast, ionic liquids with poly(hydrogen fluoride) as the anion have extremely low viscosity, and therefore have good low-temperature properties.

In the relationship between the molecular structure and conductivity of imidazolium ionic liquid, viscosity increases with increase in bulkiness of the *N*-alkyl group attached to the imidazolium ring, resulting in a drastic

Table 6.4 Electroconductivity of poly(hydrogen fluoride) salt ionic liquids

Ionic liquid	Electroconductivity (mS cm^{-1})
Me_4NF-4HF	196.6
Et_4NF-4HF	99.2
n-Pr_4NF-4HF	33.6
Et_3N-3HF	32.6
[emim]$H_{2.3}F_{3.3}$	100

decrease in electroconductivity. On the other hand, even though the *N*-alkyl group is replaced with a hydrogen atom, which is smaller than the methyl group, the viscosity increases owing to hydrogen bonding with the counter anion. The conductivity is therefore not improved. The Walden role, in which the product of mobility and viscosity is constant, can be applied to ionic liquids, and molar electroconductivity against viscosity of ionic liquids obeys the following equation:

$$\lambda\eta = \text{constant}(\lambda = \text{molar electroconductivity}, \eta = \text{viscosity})$$

Accordingly, the conductivity of ionic liquids increases with decreasing viscosity. It is reasonable that the viscosity changes depending on temperature, and the electroconductivity of ionic liquids (emim salts) drastically decreases as temperature is decreased, as shown in Figure 6.20. This is due to the rapid increase in their viscosity with decreasing temperature. This trend causes a big problem when electrochemical devices like batteries and capacitors that use ionic liquids as electrolytes are utilized during winter in a cold area. The development of ionic liquids with excellent conductivity is eagerly anticipated.

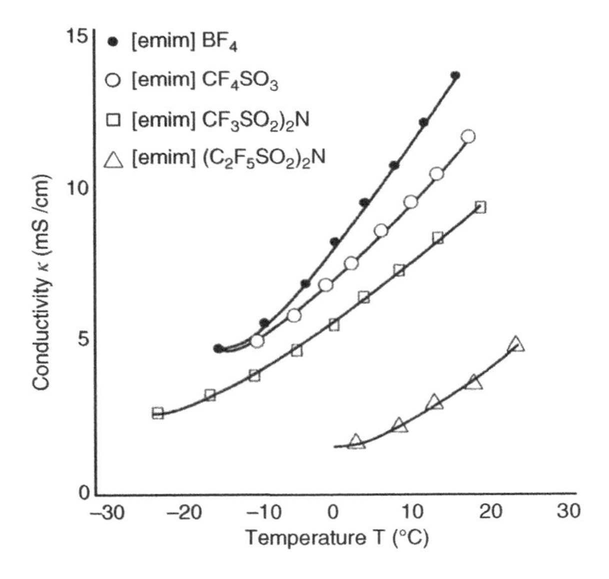

Figure 6.20 Temperature dependency of electroconductivity of imidazolinium salt ionic liquids

Ionic liquids are electrochemically stable and their usable potential region, i.e. their potential window where the ionic liquid itself can be neither oxidized nor reduced, is very large. This is one of the characteristics of ionic liquids. In general, oxidative stability depends on the anion while reductive stability depends on the cation. However, although the imidazolium ion has a positive charge, it is more easily oxidized compared to BF_4^- owing to the unsaturated bonds of the imidazole ring. Accordingly, the oxidative stability of imidazolium ionic liquids is not always attributable to the oxidative decomposition of anions. Among organic anions, trifluoroacetate ion is the most easily decomposed oxidatively, while $(CF_3SO_2)_2N^-$ and $(CF_3SO_2)_3C^-$ have oxidation resistivity and are not easily decomposed oxidatively. The order of oxidation resistivity is as follows, and this order agrees with that of oxidation potentials measured in a solution:

$$AsF_6^- > PF_6^- > BF_4^- > (CF_3SO_2)_3C^- > (CF_3SO_2)_2N^- > CF_3SO_3^- > CF_3CO_2^-$$

On the other hand, reductive stability generally depends on the reductive decomposition of the cation and the aliphatic ammonium ion is more difficult to reduce compared to the imidazolium ion. 2-Methylimidazolium ion is more difficult to reduce by 0.3–0.5 V compared a non-substituted imidazolium ion. As shown in Figure 6.21, ionic liquids consisting of an aliphatic ammonium ion and $(CF_3SO_2)_2N^-$ (TFSA) have wide potential window of about 6 V.

Phosphonium ionic liquids are superior to aliphatic ammonium ionic liquids in oxidation resistance, and they have a wide potential window (about 6.5 V) as well as thermal stability.

Poly(hydrogen fluoride) salt ionic liquids, however, have a narrow negative potential window because of the reduction of acidic proton, as shown in Figure 6.22, but such easy proton reduction assists electrochemical fluorination preferentially (see sections 5.7 and 5.8.6). Et_3N-3HF is easily oxidized because of contaminated free amine, while Et_4NF-4HF and Et_3N-5HF have excellent oxidation resistance, and their protons are readily reduced at the cathode to generate hydrogen gas.

6.8.5 Voltammetry in Ionic Liquids

When ionic liquids are used as a solvent for the measurement of the cyclic voltammetry of organic compounds, the observed redox currents are

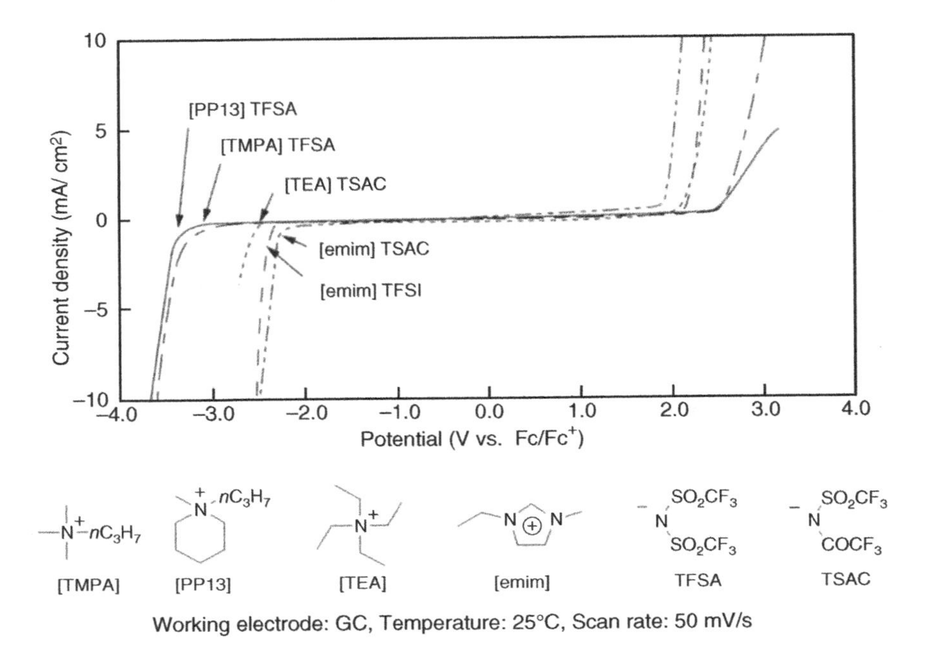

Figure 6.21 Linear sweep voltammograms in various ionic liquids

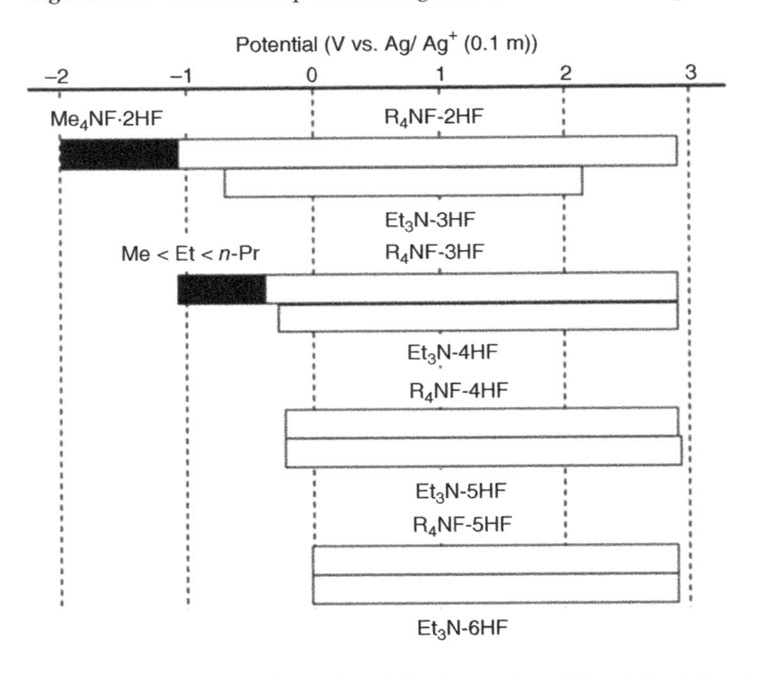

Figure 6.22 Potential windows of poly(hydrogen fluoride) salt ionic liquids

generally extremely small. This is due to the high viscosity of ionic liquids, which decreases the diffusion coefficient. For instance, the diffusion coefficient for Ni(II) salen in ionic liquid [bmim]BF_4 at room temperature is $1.8 \times 10^{-8}\,cm^2\,s^{-1}$, which is more than 500 times smaller than that in a typical organic solvent–electrolyte system like 0.1 M Et_4NClO_4/DMF [36].

6.8.6 Organic Electrochemical Reactions in Ionic Liquids

Although ionic liquids are considered to be an ideal medium for organic electrolytic reactions because of their non-flammability and sufficient electroconductivity as well as wide potential windows, there have not yet been many reports of organic electrochemical reactions in ionic liquids.

6.8.6.1 Organic Electrosynthesis

Cathodic reduction of carbonyl compounds like benzaldehyde and acetophenone has been investigated in ionic liquids, and notably their dimerization proceeded predominantly [37,38]. For acetophenone, the corresponding pinacol was formed as a diastereomeric mixture, and the diastereoselectivity is greatly affected by the ionic liquids used, as shown in Eq. 6.9 [37]. As a solvent, aliphatic ionic liquids result in higher diastereoselectivity compared to aromatic ionic liquids.

$$PhC(O)Me \xrightarrow[\text{Divided cell, 1.0 F/mol}]{-2e,\ 2H^+} Ph(Me)C(OH)\text{–}C(OH)(Me)Ph \qquad (6.9)$$

[emim]CF_3SO_3	92% yield, 6% de
[Et_3BuN]CF_3SO_3	82% yield, 58% de
[Me_3BuN]CF_3SO_3	80% yield, 54% de

Electrocatalytic homocoupling of PhBr and $PhCH_2Br$ can be carried out in the presence of NiCl(bpy) complex in [bmim]Tf_2N, as shown in Eq. 6.10 [39].

$$2\ Ph(CH_2)_xBr \xrightarrow[\text{cat. }NiCl_2(bpy),\ [bmim]Tf_2N,\ -1.4\text{ V vs Ag / Ag}^+,\ 2\text{ F / mol}]{2e} Ph(CH_2CH_2)_xPh \qquad (6.10)$$

$x = 0$: 35 %
1 : 75 %

Interestingly, Pd nanoparticles generated cathodically in ionic liquid were shown to be a highly effective ligand-free catalyst for the coupling of aryl halides [40].

Electroreductive dehalogenation of *vic*-dihalides using a Co(II)salen complex in [bmim]BF_4 has also been achieved, as shown in Eq. 6.11 [41]. The product isolation is much easier compared to the similar dehalogenation in ordinary molecular solvents since the Co(II)salen complex remains in the ionic liquid phase during product extraction with non-polar organic solvents like diethyl ether. Furthermore, the recyclability of the catalyst/ionic liquid system was demonstrated.

Br
Br
[bmim]BF_4
2 Co(I)salen 2 Co(II)salen
2e
68%
(6.11)

Cyclic carbonates are prepared by the reduction of CO_2 at −2.4 V vs. Ag/AgCl in the presence of epoxides in various ionic liquids like [emim]BF_4, [bmim]PF_6 and [BPy]BF_4 (BPy=*n*-butylpyridinium) using a Cu cathode and a Mg or Al anode [42].

Electrochemical synthesis of carbamates was also achieved by the electrolysis of a solution of CO_2 and amine in [bmim]BF_4 followed by the addition of alkylating agent, as shown in Eq. 6.12 [43].

Bn
N–H
Me
−e
[bmim]BF_4
CO_2 (0.1 MPa), 55°C
Divided cell
−2.4 V vs. Ag/AgCl
then EtI
Bn
N
Me
O
OEt
87%
(6.12)

Furthermore, a kinetic study on the anodic coupling of aromatic compounds has been carried out in ionic liquids. Anodic oxidation of 1,2-dimethoxybenzene led to the corresponding trimer, as shown in Eq. 6.13 [44]. In this reaction, the dimerization rate in ionic liquid is five to ten times smaller than that in acetonitrile because of the high viscosity of the ionic liquid.

OMe OMe $\xrightarrow[\text{[emim]TFSA}]{-6e,\ -6H^+}$ OMe OMe, 3 (6.13)

N-Heterocyclic carbenes can be generated by cathodic reduction of imidazolinium-based ionic liquids [45–47]. The resulting carbenes are stable bases that are able to catalyze the Henry reaction as shown in Eq. 6.14 [47].

BF_4^- Me–N⊕N–Bu, H $\xrightarrow[\text{[bmim]}BF_4]{e}$ Me–N N–Bu + $\frac{1}{2}$ H_2

Ph H (O) + $MeNO_2$ $\xrightarrow[\text{Me–N N–Bu}]{}$ HO NO_2 Ph 81% + NO_2 Ph 17% (6.14)

Since ionic liquids generally have much higher viscosities, mass transport is quite slow, as described before. This is a disadvantage for electrosynthesis in ionic liquids. However, it was found that cathodic reduction of *N*-methylphthalimide was promoted under ultrasonication, resulting higher conversion and current efficiency, as shown in Eq. 6.15 [48]. This is due to facilitated mass transport of the substrate under ultrasonication.

O NH O $\xrightarrow[\text{[emim]}NTf_2\ \ 0.11\ F/mol]{2e^-,\ 2H^+}$ OH NH O (6.15)

Sonication: 89% (current efficiency)
Non-sonication: 66% (current efficiency)

6.8.6.2 Electrochemical Fluorination

Poly(hydrogen fluoride) salts consisting of amine and ammonium fluoride and hydrogen fluoride (Et_3N-nHF, $n = 3$–5; Et_4NF-nHF, $n = 3$–5) are ionic liquids that have good conductivity because of their low viscosity [49,50]. They become more anodically stable with increasing HF content (increasing n). As described in Chapter 5, conventional electrochemical fluorination has been carried out in organic solvents containing fluoride salts, such as Et_3N-3HF and Et_4NF-3HF [49–51]. However, anodic passivation (the formation of a non-conducting polymer film on the electrode surface that suppresses Faradaic current) takes place often, resulting in poor yield and low current efficiency. Moreover, acetoamidation also occurs preferentially when MeCN is used as an electrolytic solvent. In order to overcome such problems, solvent-free electrochemical fluorination is an alternative method of preventing anode passivation and acetoamidation. Solvent-free selective electrochemical fluorination of benzenes, naphthalene, olefins, furan, benzofuran and phenanthroline was first achieved in less than 50% yield using the ionic liquid Et_3N–3HF as the reaction medium, supporting electrolyte and a fluorine source [52].

Although Et_3N-3HF and Et_4NF-3HF are easily oxidized (lower or around 2 V vs. Ag/Ag^+), anodically very stable poly(hydrogen fluoride) salts, Et_4NF-mHF ($m > 3.5$) and Et_3N-5HF (3 V vs. Ag/Ag^+) have also been developed [50]. Using ionic liquid Et_4NF-mHF ($m > 3.5$), solvent-free anodic partial fluorination of arenes such as various substituted benzenes, toluene and quinolines was successfully carried out at high current densities with good current efficiencies (66–90%) [53,54]. As shown in Eq. 6.16, monofluorotoluene and difluorotoluene are formed successively with increasing electricity, however, even when the electricity is increased further trifluorotoluene is not formed and ring fluorination starts [54].

$$C_6H_5CH_3 \xrightarrow[\substack{Et_4NF\text{-}4HF \\ 2.0\ F/mol \\ 1.9\ V\ vs.\ Ag/Ag^+}]{-2e,\ -H^+} C_6H_5CH_2F\ (52\%) \xrightarrow[\substack{Et_4NF\text{-}4HF \\ 2.0\ F/mol \\ 2.1\ V}]{-2e,\ -H^+} C_6H_5CHF_2\ (47\%) \xrightarrow[\substack{Et_4NF\text{-}4HF \\ 2.0\ F/mol \\ 2.5\ V}]{-2e,\ -H^+} F\text{-}C_6H_4CHF_2 \quad (6.16)$$

Selective electrochemical formyl hydrogen-exchange fluorination of aliphatic aldehydes affords acyl fluorides using Et_3N–5HF [55]. Selective anodic fluorination of cyclic unsaturated esters in Et_3N–5HF is also accomplished to provide ring-expansion fluorinated products, as shown in Eq. 6.17 [56,57].

CO_2Et —(−2e; Et_3N-5HF; 2.2–2.3 V vs. Ag/Ag^+; −20°C)→ F F CO_2Et (6.17)

n = 1 71%
n = 0 56%

Furthermore, anodic fluorination of various phenol derivatives can be performed in Et_3N-5HF using carbon fibre cloth as an anode to form 4,4-difluorocyclohexadienone derivatives in good yields (Eq. 6.18) [58].

OH —(−4e, $-2H^+$; Et_3N-5HF; Carbon fibre cloth)→ O, F F (6.18)

89%

Electrochemical fluorination of adamantanes is also possible in Et_3N–5HF. Mono-, di-, tri- and tetrafluoroadamantanes were selectively prepared from adamantanes by controlling oxidation potentials, and the fluorine atoms were introduced selectively at the tertiary carbons, as shown in Eq. 6.19 [59].

—(−2e, $-H^+$; Et_3N-5HF/CH_2Cl_2; 2.3–3.0 V (vs. Ag/Ag^+))→ F + F, F + F, F, F + F, F, F, F

(6.19)

Anodic fluorination of cyclic ethers, lactones and open-chain and cyclic carbonates can be achieved by anodic oxidation of a mixture of a large amount of the liquid substrate and a small amount of Et_4NF–4HF (only 1.5–1.7 equiv. of F^- to the substrate) at a high current density (150 mA cm^{-2}) (Eqs. 6.20 and 6.21) [60]. In this method, the substrate was selectively oxidized to provide the corresponding monofluorinated product in good yield and with good current efficiency. In sharp contrast, the use of organic solvents or a large amount of Et_4NF-4HF resulted in no formation or low yield (about 10%) of the desired fluorinated product. Isolation of the fluorinated products is easy: fluorinated lactone and carbonates can be isolated by extraction with solvent, while fluorinated tetrahydrofuran can be easily isolated by distillation of the reaction mixture after electrolysis. In these cases, the substrates are predominantly oxidized to result in efficient fluorination because only a small amount of supporting fluoride salt is used. Since platinum with a low hydrogen overpotential is used as a cathode, the acidic protons of the ionic fluoride salt are predominantly reduced at the cathode to generate hydrogen gas, hence a separator for the electrolysis is not necessary. As explained, a small amount of fluoride salt is enough for the fluorination, therefore this fluorination method is desirable from an atom economical perspective.

$-2e, -H^+$; Et_3N-4HF

$X = CH_2, n = 0$ 83%
$X = O, n = 1$ 80%
$X = O, n = 0$ 77%

(6.20)

X–CO–OCH_2CH_3 → X–CO–$OCHFCH_3$; $-2e, -H^+$; Et_4NF-5HF; 2–2.5 F/mol

X = EtO 97%
X = Et 44%

$-2e, -H^+$; Et_4NF-5HF

$Z = CH_2$ 75%
Z = O 87%

(6.21)

Since phthalide is hardly oxidized ($E_p^{ox} = 2.81$ V vs. SCE), anodic fluorination does not proceed in either organic solvent or a solvent-free system. However, fluorination takes place highly efficiently in a double ionic liquid system consisting of [emim]OTf and Et_3N-5HF, as shown in Eq. 6.22 [61]. In this double ionic system, the cationic intermediate generated from the phthalide was expected to have a TfO^- counter anion (activated cation **A** in Eq. 6.22), which readily reacted with F^- to provide the fluorinated phthalide in good to excellent yield.

$$\text{phthalide} \xrightarrow[\substack{[\text{emim}][\text{TfO}] \\ n = 3 \text{ or } 5}]{-2e,\ -H^+ \quad Et_3N\text{-}nHF} \left[\mathbf{A}\ (\text{activated cation},\ {}^{\ominus}OTf)\right] \xrightarrow{F^-} \text{3-fluorophthalide} \quad n = 3\ \ 78\%,\ n = 5\ \ 90\% \tag{6.22}$$

The viscosity of ionic liquid fluoride salts is higher than that of ordinary organic solvents, therefore the mass transport of substrates to the anode surface from the bulk liquid is slower than in organic solvents, which is unfavourable for electrolytic reactions.

As mentioned in section 6.11, it is known that ultrasonication greatly enhances mass transport from bulk to electrode surface. This effect can also be applied to an ionic liquid system as the efficiency of anodic fluorination in ionic liquid fluoride salts is markedly increased, as shown in Eq. 6.23 [62]. Notably, anodic difluorination of ethyl α-(phenylthio) acetate proceeds efficiently even in readily oxidizable ionic liquid, Et_3N-3HF, which is not suitable for difluorination.

$$EtOC(O)CH_2SPh \xrightarrow[\substack{\text{Under ultrasonication} \\ \text{Yield: } 65\%}]{-4e,\ -2H^+ \quad 6\ \text{F/mol}\ \ Et_3N\text{-}3HF} EtOC(O)CF_2SPh$$
$$EtOC(O)CH_2SPh \xrightarrow[\substack{Et_3N\text{-}3HF/MeCN \\ 1.6\ \text{V vs SSCE} \\ 2.5\ \text{F/mol} \\ \text{Yield: } 76\%}]{-2e,\ -H^+} EtOC(O)CHFSPh \xrightarrow[\substack{Et_3N\text{-}3HF/MeCN \\ 2.2\ \text{V vs SSCE} \\ 20.7\ \text{F/mol} \\ \text{Yield: } 53\%}]{-2e,\ -H^+} EtOC(O)CF_2SPh \tag{6.23}$$

In neat ionic liquid fluoride salts, the nucleophilicity of fluoride ions is rather low, resulting in poor fluorination yields. It has been demonstrated that ether solvents like DME enhance the nucleophilicity of fluoride ions, but DME is rather easily decomposed [63]. In contrast, PEG and even its oligomer are stable against anodic oxidation, and it was found that the addition of only about 3% PEG oligomer to the reaction system greatly improved the yield due to its ability to coordinate the counter cations of fluoride ions, as shown Eq. 6.24 [64].

−4e
Et_3N-3HF
8 F/mol
50 mA/cm^2

Without additive: 22%
With DME: 18%
With PEG (MW 200): 80%

(6.24)

Severe passivation of the anode often occurs even in ionic liquid HF salts. Use of mediators is effective to solve the problem, but product separation becomes complicated. Fuchigami and co-workers have developed a novel indirect anodic fluorination system employing a task-specific ionic liquid with an iodoarene moiety as the mediator in HF salts [65]. The mediator improved the reaction efficiency for a variety of electrochemical fluorinations (Figure 6.23) and remained intact in the ionic liquids after the product extraction process for reuse in subsequent runs.

In addition, a polymer-supported iodobenzene (PSIB) mediator is also effective for indirect anodic fluorination in HF salts [66]. In this case, the iodobenzene moiety pendent from the solid polymer support cannot be directly oxidized, therefore a double mediator system is necessary. As shown in Figure 6.24, anodic oxidation of Cl^- gave Cl^+, which reacted with the iodobenzene moiety to form PhI^+Cl. This species then captures a fluoride ion to give the hypervalent [(chloro)(fluoro)iodo] benzene moiety. The hypervalent iodine moiety thus generated oxidizes the substrate and

−2e, 2F$^-$
F_2I–Ph– I–Ph–
Et_3N-3HF
5 mA/cm^2
4 F/mol

EWG = COOEt: 87%
(Without mediator: 31%
EWG = CN: 72%

I–Ph– = I–C_6H_4–O$(CH_2)_3$–N$^+$ (imidazolium) $(CF_3SO_2)_2N^-$

Figure 6.23 Electrochemical fluorination using task-specific ionic liquid with mediator

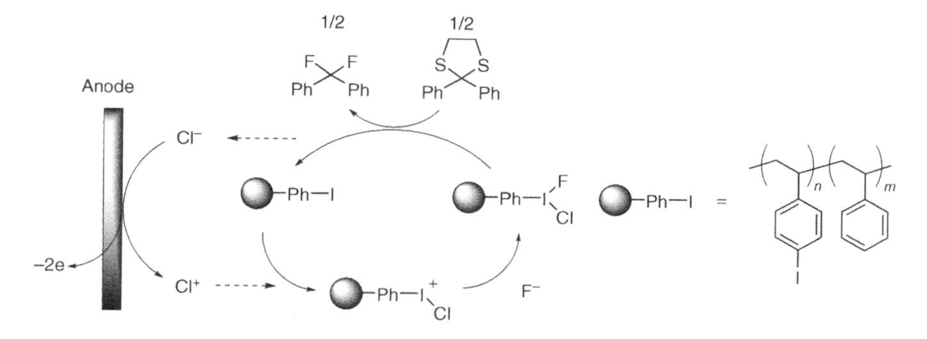

Figure 6.24 Electrochemical fluorination using a polymer-supported mediator

consequently the starting PSIB is recovered. The recovered PSIB mediator is reused in subsequent runs, maintaining a good yield (about 90%) of the fluorinated product.

6.8.6.3 Electropolymerization

Mattes's group and Fuchigami's group independently achieved the electrooxidative polymerization of pyrrole, thiophene and aniline in different ionic liquids [67–69]. Mattes's group used 1-butyl-3-methylimidazolium tetrafluoroborate and hexafluorophosphate ($[bmim]BF_4$ and $[bmim]PF_6$) for electropolymerization of pyrrole and aniline. They found that the π-conjugated polymers thus obtained are highly stable and can be electrochemically cycled in ionic liquids up to a million times [67]. In addition, because the polymers have cycle-switching speeds as fast as 100 ms, they can be used as electrochromic windows and numeric displays. Furthermore, it was demonstrated that polyaniline prepared in this way is highly useful for electrochemical actuators. A 10-mm length of 59-μm diameter wet-spun polyaniline emeraldine base (EB) fibre is treated with trifluoromethanesulfonic (triflic) acid to form the corresponding emeraldine salt (ES) fibre, which has high conductivity of $300\,S\,cm^{-1}$. The fibre with ES structure is reduced with two electrons at about −0.4 V vs. Ag/Ag^+ in ionic liquid $[bmim]BF_4$ to form leucoemeraldine (LE), and LE is oxidized at about +0.8 V vs. Ag/Ag^+ to form original ES, as shown in Figure 6.25. Thus, by oxidation and reduction, the cation of the ionic liquid, [bmim], is expelled and incorporated, respectively. The strain is therefore contractile when the polymer is oxidized from the fully reduced LE state to ES, and expansive upon reduction (Figure 6.25). In lifetime tests of the fibre, both

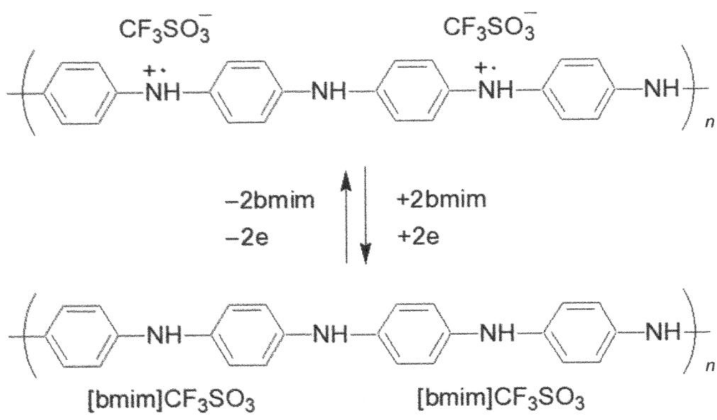

Figure 6.25 Principle of an electrochemical actuator

electroactivity and electromechanical actuation continue without significant decrease (<1%) in either stress or strain for 10,000 redox cycles because anhydrous ionic liquid is used as a solvent. The π-conjugated polymer in an ionic liquid electrolyte system is therefore highly promising for electrochemical mechanical actuators.

Fuchigami's group employed [emim]OTf for electrochemical polymerization [68,69]. They found that the polymerization of pyrrole in ionic liquid proceeds much faster than that in conventional media like aqueous and acetonitrile solutions containing 0.1 M [emim]OTf as a supporting electrolyte. Interestingly, as shown in Table 6.5, the surface of the polypyrrole film prepared in neat [emim]OTf is so smooth that no grains

Table 6.5 Physical properties of polypyrrol and polythiophene films prepared electrochemically in various media

Polymer	Media	Roughness factor[a] (dimensionless)	Electro-chemical capacitance ($C\ cm^{-3}$)	Electroconductivity ($S\ cm^{-1}$)	Dopin level (%)
Polypyrrole	H_2O	3.4	77	1.4×10^{-7}	22
Polypyrrole	CH_3CN	0.48	190	1.1×10^{-6}	29
Polypyrrole	[emim]OTf	0.29	250	7.2×10^{-2}	42
Polythiophene	CH_3CN	8.6	9	4.1×10^{-8}	–
Polythiophene	[emim]OTf	3.3	45	1.9×10^{-5}	–

[a] Standard deviation of thickness.

are observed, and both the electrochemical capacity and the electroconductivity are markedly increased when the polypyrrole and polythiophene films are prepared in the ionic liquid. This may be attributable to the extremely high concentration of anions as dopants, which results in a much higher doping level. As described above, polymer films prepared in the ionic liquid have a higher electrochemical density and highly regulated morphological structures, and therefore they have possible uses as high-performance electrochemical capacitors, ion-sieving films, ion-selective electrodes, matrices for hosting catalyst particles and so on.

Anodic polymerization of 3-(*p*-fluorophenyl)thiophene was also carried out in [emim]NTf_2 and the resulting polymer film was also found to be very smooth [70]. Furthermore, electrosynthesis of poly(3,4-ethylenedioxythiophene) (PEDOT) and polyphenylene in ionic liquids has been reported [71,72].

An ionic liquid is a recyclable medium for organic synthesis, which is one of the pronounced characteristics of ionic liquids. However, when an ionic liquid is used as electrolytic medium, decomposition of the ionic liquid itself often occurs at a counter electrode. This is a big problem when the ionic liquid is recycled. In the case of electrochemical polymerization, however, this problem can be solved using cyclic potential-scanning oxidative polymerization. In this polymerization method, monomer is oxidized to form polymer film, which deposits on the electrode. Additionally, the deposited polymer itself is oxidized and reduced repeatedly during alternately anodic and cathodic scanning. During reduction at the working electrode, anodic oxidative polymerization of monomer and oxidation of the polymer itself (so-called doping) therefore occur at the counter electrode, while during oxidation at the working electrode, reduction of the polymer (so-called dedoping) occurs at the counter electrode, hence decomposition of the ionic liquid does not occur. Ionic liquid is therefore easily recovered simply by extracting the remaining monomer with appropriate solvents, and recovered ionic liquid can be used many times [68,69].

6.8.6.4 *Others*

Although the following synthetic application of ionic liquids is not for organic substances, electrochemical deposition of various metals in ionic liquids is possible. For instance, Zn, Ga and In can be electrochemically deposited from mixtures of [emim]Cl and $ZnCl_2$, $GaCl_3$ and $InCl_3$, respectively. A variety of typical elements and transition metals can be deposited from ionic liquids other than chloride ionic liquid. Furthermore, it is known that one-electron electrochemical reduction of oxygen molecules in ionic liquids generates superoxide ions efficiently.

6.9 THIN-LAYER ELECTROLYTIC CELLS

Although electro-organic syntheses have been established as a powerful tool in organic syntheses, they still are in development to fulfil their potential as a 'green' methodology. In conventional electrosynthetic processes, a large amount of supporting electrolyte has to be added to the solvent to give sufficient electrical conductivity, but the presence of a large amount of the supporting electrolyte might cause separation problems in the reaction mixture work-up and additional waste problems.

The capillary gap cell developed at BASF for the commercial electrosynthetic process (paired electrosynthesis of phthalide and *t*-butylbenzaldehyde) allows the electrolysis to be conducted even in a dilute electrolyte solution. The cell consists of circular disk electrodes with a small interelectrode gap (1–2 mm) to minimize the ohmic voltage drop in the electrolyte (Figure 6.26) [73].

On the other hand, in laboratory-scale experiments, the use of a simple thin-layer flow-cell geometry with working and auxiliary electrodes directly facing each other allows electrosynthetic processes to be conducted in

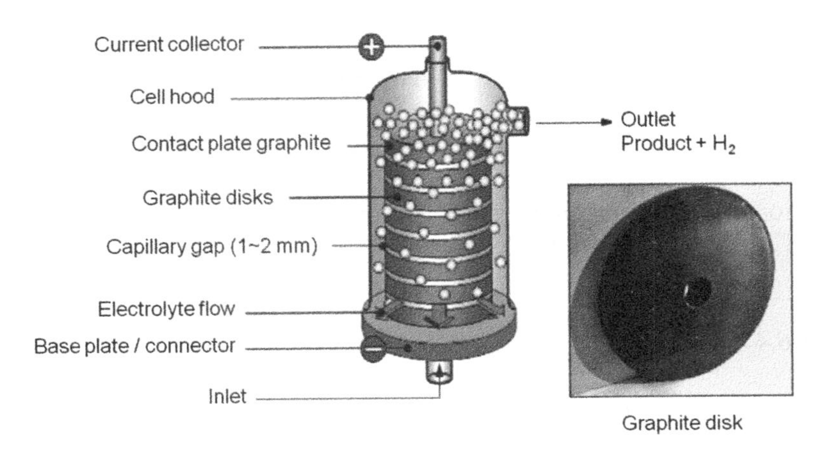

Figure 6.26 Schematic representation of the capillary gap electrochemical cell of BASF SE

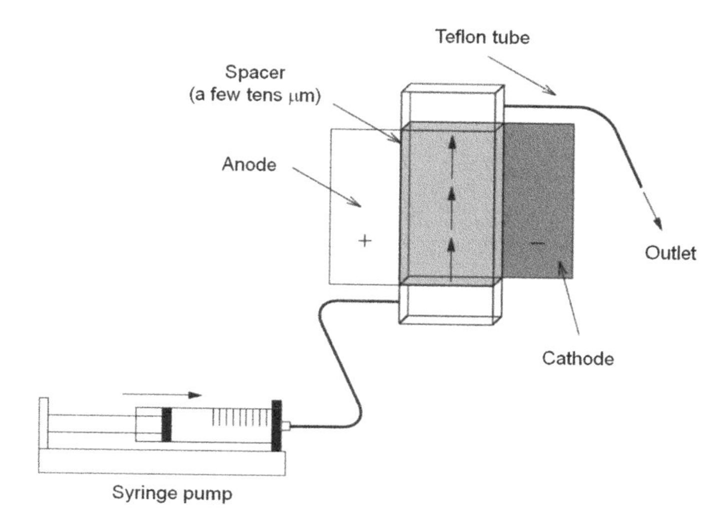

Figure 6.27 Schematic representation of a thin-layer flow cell

flow-through mode in the absence of supporting electrolyte (Figure 6.27) [74]. For example, by using this type of cell, the methoxylation of furan is conducted anodically in methanolic solution even in the absence of added supporting electrolyte [75].

6.10 ELECTROCHEMICAL MICROFLOW SYSTEMS

Recently, microflow systems have received significant research interest from both academia and industry. The fundamental advantages and potential benefits of microflow technology are (i) an extremely large surface-to-volume ratio, (ii) precise temperature control, (iii) precise residence time control, (iv) strict laminar flow control, (v) extremely fast molecular diffusion and (vi) improvement of reaction process safety. As introduced above, microflow systems have many advantages that can be applied in a wide range of organic synthesis and organic mass production processes. Electron transfer is one of the most common driving forces for organic reactions, and organic electrosynthesis serves as a straightforward and powerful method of organic electron-transfer processes. The integrated use of microflow technology with organic electrosynthesis is one of the most sophisticated processes in organic chemistry. In addition, novel systems that are realized only by using electrochemical microflow reactors have been developed.

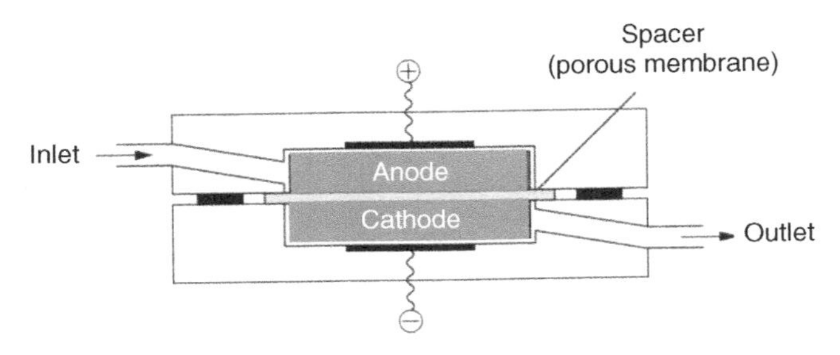

Figure 6.28 Schematic representation of a microflow electrochemical reactor

Yoshida *et al.* reported that a microflow electrochemical system serves as a quite effective method for oxidative generation of unstable organic cations at low temperatures [76]. This method is called the cation-flow method. An electrochemical reactor for the cation-flow method is equipped with a carbon felt anode and a platinum wire cathode (Figure 6.28). The anodic chamber and the cathodic chamber are separated by a diaphragm of PTFE membrane. A solution of a cation precursor is introduced to the anodic chamber and a solution of trifluoromethanesulfonic acid (TfOH) as a proton source is introduced to the cathodic chamber. The organic cation that is generated is immediately transferred to a vessel in which a nucleophilic reaction takes place to give the desired coupling product.

Atobe and co-workers reported that the use of parallel laminar flow in a microflow electrochemical reactor enables the effective generation of an *N*-acyliminium ion followed by trapping with an easily oxidizable carbon nucleophile such as allyltrimethylsilane (Figure 6.29) [77]. A solution of a cation precursor and a solution of allyltrimethylsilane are introduced in a parallel manner. The laminar flow prevents the oxidation of allyltrimethylsilane at the anode. Only the precursor is oxidized to generate the *N*-acyliminium ion. The *N*–acyliminium ion that is generated diffuses and reacts with allyltrimethylsilane. Although the efficiency of the process is very low for the Bu_4NBF_4/CH_3CN system, use of the Bu_4NBF_4/TFE (2,2,2-trifluoroethanol) system (59% yield) or an ionic liquid (62–91% yield) gave rise to the formation of the desired product.

The use of parallel laminar flow in a microflow electrochemical reactor also enables chemoselective cathodic reduction to control product regioselectivity in electrochemical carbonyl allylation [78]. Electrochemical carbonyl allylation can produce either γ- or α-adducts depending on whether the aldehyde or allylic halide is reduced by the cathode. If the

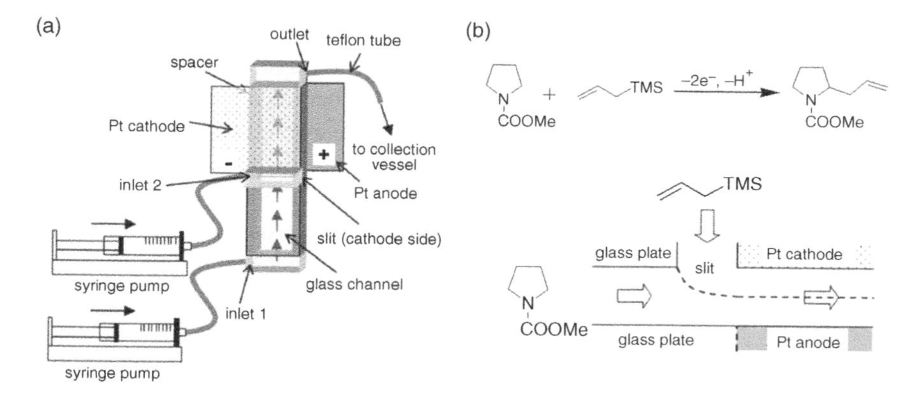

Figure 6.29 Schematic representation of (a) a two-inlet microflow reactor system and (b) parallel laminar flow in the reactor. The illustrated model reaction is anodic substitution reaction of *N*-(methoxycarbonyl)pyrrolidine with allyltrimethylsilane

aldehyde has a higher reduction potential, the allylic halide is predominantly reduced to give the γ-adduct, but if the reduction potential of the allylic halide is higher, the aldehyde is reduced preferentially and the α-adduct generation is favoured. Control of product selectivity (regioselectivity in this reaction) therefore requires that either the allylic halides or the aldehydes should be reduced chemoselectively, regardless of their reduction potentials (Eq. 6.25). To perform chemoselective cathodic reduction, the authors employed a liquid–liquid parallel laminar flow formed in an electrochemical microflow reactor. As shown in Figure 6.30,

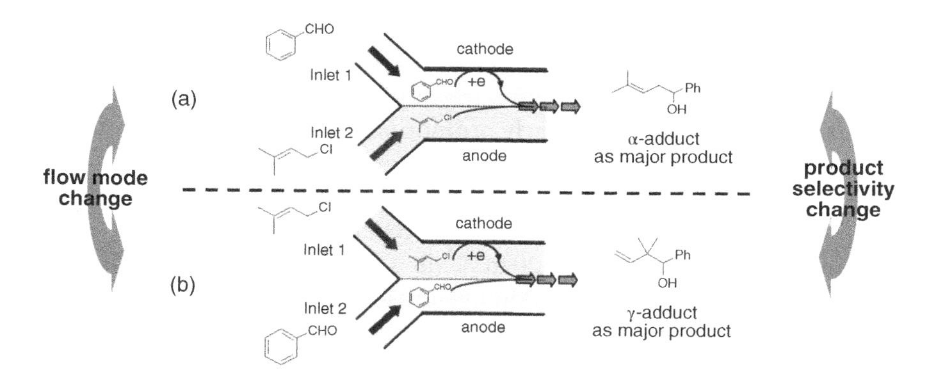

Figure 6.30 Chemoselective cathodic reduction using parallel laminar flow in a two-inlet microflow reactor. (a) Flow mode for the selective reduction of benzaldehyde. (b) Flow mode for the selective reduction of 1-chloro-3-methyl-2-butene

when two solutions (allylic chloride solution and aldehyde solution) are introduced through inlets 1 and 2 of the microflow reactor, a stable liquid–liquid interface is formed and mass transfer between the input streams occurs only by means of diffusion. The substrate introduced through inlet 1 can therefore be predominantly reduced, whereas the reduction of the inlet 2 substrate (inflow 2) can be avoided. Consequently, chemoselective cathodic reduction proceeds and an intentional cross-coupling product is obtained regioselectively. In other words, product selectivity control is realized by simply switching the reagent flows.

γ α Cl + PhCHO $\xrightarrow[2H^+, -Cl^-]{2e}$ Ph OH (γ-adduct), Ph OH (α-adduct) (6.25)

6.11 ELECTROLYSIS UNDER ULTRASONICATION

The many benefits of ultrasound in chemical processes are well known and have been investigated in a variety of chemical fields, but perhaps the most striking influence of ultrasound concerns heterogeneous reaction systems, particularly those with a solid–liquid interface where particle size modification, the modification of particle dispersion, the enhancement of mass transport, the cleaning of surfaces or the formation of fresh surfaces are among the beneficial processes [79].

Since an electrochemical synthetic process is a typical heterogeneous one in a solid (electrode)–liquid (electrolytic solution) interface, various effects of ultrasound, particularly promotion of mass transport, would be induced by ultrasonication. A major contribution to mass transport is the micro-jet stream resulting from the asymmetric collapse of a cavitation bubble (Figure 6.31) [80]. Suslick *et al.* reported that the velocity of the stream reaches in excess of 100 m s^{-1} in a water–solid interface [81].

Such a mass transfer promotion by ultrasonication provides an increase in the current efficiency for a variety of electrosyntheses. For example, Atobe *et al.* reported that a significant ultrasonic effect on current efficiency was found in the electrochemical reduction of *p*-methylbenzalaldehyde (Eq. 6.26) [82]. As shown in Table 6.6, current efficiency for the reduction of *p*-methylbenzalaldehyde was dramatically increased under

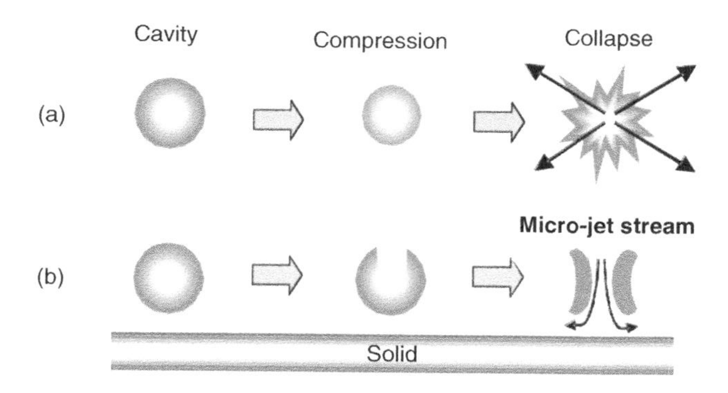

Figure 6.31 Collapse of cavitation bubble. (a) Symmetrical collapse of a cavitation bubble in bulk solution. (b) Asymmetrical collapse of a cavitation bubble in a liquid–solid interface

ultrasonication. Furthermore, product selectivity for the hydrodimeric product (D1) was also increased by ultrasonication, and the effects could be rationalized experimentally and theoretically as being due to the promotion of the mass transport of the substrate molecule to the electrode surface from the electrolytic solution by ultrasonic cavitation [83,84].

Table 6.6 Electroreduction of *p*-methylbenzaldehyde, dimethyl maleate and benzoic acid

Starting compound	Stirring mode	Current efficiency for [D1] + [M1], [D2] + [M2] or [M3] + [M4]/%	Selectivity [D1]/[M1], [D2]/[M2] or [M3]/[M4]
p-methylbenzaldehyde[a]	Still	35	0.0
p-methylbenzaldehyde[a]	Mechanical[b]	74	1.0
p-methylbenzaldehyde[a]	Ultrasonic[c]	89	24
Dimethyl maleate[d]	Still	66	0.0
Dimethyl maleate[d]	Mechanical[b]	93	0.3
Dimethyl maleate[d]	Ultrasonic[c]	96	0.4
Benzoic acid[e]	Still	1	0.0
Benzoic acid[e]	Mechanical[b]	17	0.2
Benzoic acid[e]	Ultrasonic[c]	54	1.5

[a] Electrolyzed by passing 0.5 F mol^{-1} at 20 mA cm^{-2} on a lead cathode in a 0.25 M H_2SO_4/50% MeOH solution.
[b] Stirred by a rotating magnet bar.
[c] The cathode was positioned 1.7 cm apart from the top of an ultrasonic horn (0.6 cm diameter, 12 W, 20 kHz).
[d] Electrolyzed at a lead cathode in a 0.025 M KH_2PO_4/0.025 M Na_2HPO_4/0.5 M NaCl solution.
[e] Electrolyzed at a lead cathode in a 0.05 M H_2SO_4/0.2 M citric acid solution.

The product selectivity in the electrochemical reduction of dimethyl maleate (Eq. 6.27) and benzoic acid (Eq. 6.28) was also greatly influenced by ultrasonication, as shown in Table 6.6.

$$\mathrm{ArCHO}\ (\mathrm{Ar}: p\text{-}CH_3C_6H_4) \begin{cases} \xrightarrow{e,\ H^+} \frac{1}{2}\ \mathrm{ArCH(OH)\text{-}CH(OH)Ar}\ [D1] \\ \xrightarrow{2e,\ 2H^+} \mathrm{ArCH_2OH}\ [M1] \end{cases} \quad (6.26)$$

$$\begin{matrix}\mathrm{CHCOOCH_3}\\ \| \\ \mathrm{CHCOOCH_3}\end{matrix} \begin{cases} \xrightarrow{e,\ H^+} \frac{1}{2}\ \mathrm{CH_2COOCH_3\text{-}CHCOOCH_3\text{-}CHCOOCH_3\text{-}CH_2COOCH_3}\ [D2] \\ \xrightarrow{2e,\ 2H^+} \mathrm{CH_2COOCH_3\text{-}CH_2COOCH_3}\ [M2] \end{cases} \quad (6.27)$$

$$\mathrm{PhCOOH} \xrightarrow[-H_2O]{2e,\ 2H^+} \underset{[M3]}{\mathrm{PhCHO}} \xrightarrow{2e,\ 2H^+} \underset{[M4]}{\mathrm{PhCH_2OH}} \quad (6.28)$$

As mentioned in section 5.8, conducting polymers exhibit not only electroconductivity but also unique optical and chemical properties [85]. The diversity of properties exhibited by conducting polymers offers these materials to be used in numerous technological applications. Generally, the properties of polymers originate from their chemical (molecular) and physical (morphological) structures. It therefore follows that the structures of conducting polymers should be able to be controlled in order to tailor them to the purposes of their utilization. Their chemical structures can be controlled by changing the molecular structures of the corresponding monomers and by selecting conditions and procedures for polymerization [85]. On the other hand, methods for controlling their physical structures have been relatively limited, but recently many studies have focused on applying ultrasound to polymerization processes, particularly electropolymerization, for this purpose.

Osawa *et al.* found that the quality of polythiophene films electropolymerized on an anode can be enhanced by ultrasound. By conventional methodology the films become brittle, but by using ultrasound from a 45-kHz cleaning bath, flexible and tough films (tensile modulus 3.2 GPa and strength 90 MPa) can be obtained [86].

The work of Atobe and co-workers was probably the first 'modern' study investigating electropolymerization under sonication in a complete series of papers at low frequencies. Starting from electro-organic reactions under ultrasonic fields [87], polymerization of aniline was studied both in electrochemical [88] and chemical routes [89,90] as well as synthesis of nanoparticle synthesis [91,92].

The behaviour of polypyrrole films electropolymerized under ultrasonication was also investigated, and unique properties in the doping–undoping processes were highlighted. The authors attributed their results to the elaboration of highly dense film under sonication, but also deplored the degradation of the film due to high cavitation at 20 kHz (Figure 6.32) [93]. The Besançon group also studied the use of high-frequency ultrasound (500 kHz, 25 W) for electropolymerization of 3,4-ethylenedioxythiophene (EDOT) or polypyrrole in aqueous medium in order to investigate its effects on conducting polymer properties. They showed that (i) mass transfer enhancement induced by sonication improves electropolymerization and (ii) the mass transfer effect is not the only phenomenon induced by ultrasound during electrodeposition [94,95].

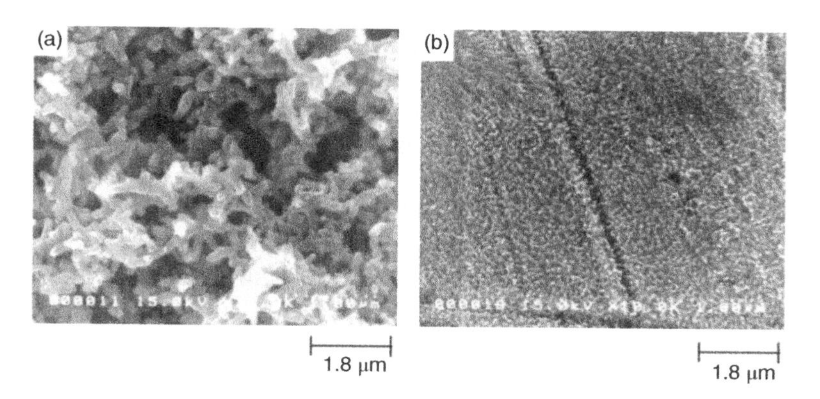

Figure 6.32 SEM images of polyaniline films prepared (a) without and (b) with ultrasonication

6.12 ELECTROSYNTHESIS USING SPECIFIC ELECTRODE MATERIALS

6.12.1 Electrochemical Synthesis Using Hydrophobic Electrodes

6.12.1.1 Hydrophobic Composite-Plated Electrodes

Most electrodes are hydrophilic, but hydrophobic electrodes can be prepared by composite-plating electrodes like Ni, Zn, Pb etc. in the corresponding metal salt-poly(tetrafluoroethylene) (PTFE) dispersion plating bath. Graphite fluoride and fluorinated pitch other than PTFE are also used as hydrophobic composite materials. During plating, the fine hydrophobic particles are incorporated into the plated layer. The electrodes thus prepared have good electroconductivity and their plated layer surface shows water-repellent properties as well as mechanical and chemical stability. In general, hydrogen and oxygen evolution occur competitively in electrolytic reactions of organic compounds in an aqueous solution. In particular, hydrogen evolution readily occurs in an acidic aqueous solution while oxygen evolution easily takes place in an alkaline aqueous solution. Accordingly, it is not easy to cathodically reduce carbonyl compounds in aqueous acidic solution and anodically oxidize alcohols in aqueous alkaline solution. High hydrogen-overpotential cathodes and oxygen-overpotential anodes therefore have to be used for electrochemical reactions in aqueous acidic and alkaline solutions, respectively. However, nickel/PTFE composite-plated Ni electrodes are used for the reduction and oxidation, as mentioned above, to provide the corresponding alcohols and carbonyl compounds, respectively, with much higher current efficiency compared to unplated Ni electrodes [96,97]. This is not because of the higher hydrogen- and oxygen-overpotentials of the composite-plated electrode suppressing hydrogen and oxygen evolution but because of the substrate-collecting effect as a result of strong hydrophobic interaction between the hydrophobic electrode surface and hydrophobic organic substrates [96,97].

$$\text{cyclopentanol (OH)} \xrightarrow[\text{1M KOH/80\% aq. MeCN}]{-2e,\ -2H^+} \text{cyclopentanone (=O)} \tag{6.29}$$

PTFE/Ni cathode: 98% current efficiency
Ni cathode: 2% current efficiency

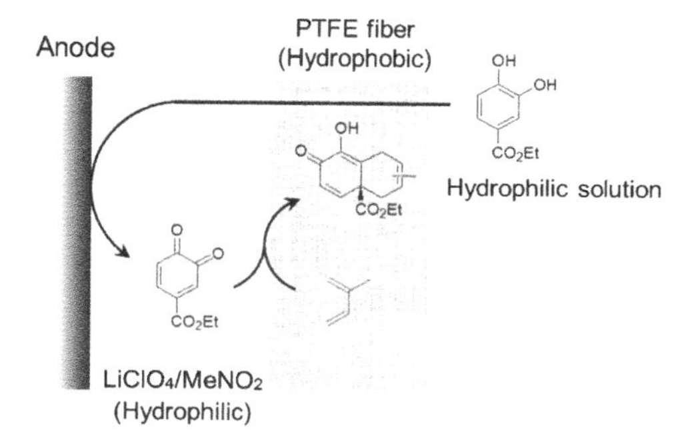

Figure 6.33 Diels–Alder reaction of anodically generated quinone derivatives with diene on the electrode surface modified with PTFE fibres

6.12.1.2 PTFE-Fibre-Coated Electrodes

PTFE-fibre-coated electrodes, which are prepared by wrapping the electrode in PTFE strings, are hydrophobic. It has been demonstrated that anodic oxidation of hydroquinones at this electrode in the presence of dienes in $NaClO_4/MeNO_2$ generated quinones, which underwent Diels–Alder reaction with the dienes to provide cyclic products in excellent yields [98]. Use of an uncoated glassy carbon anode also generates quinones from hydroquinones, but the coexisting dienes are easily oxidized and decompose prior to reaction with the quinones. However, when a PTFE-fibre-coated glassy carbon electrode is used, the easily oxidizable diene is maintained on the PTFE-fibres and only the polar hydroquinones can reach the electrode through the hydrophobic fibres, as shown in Figure 6.33. Next, the anodically generated hydrophobic quinones immediately react with the dienes adsorbed on the PTFE fibres, and the resulting hydrophobic products are also maintained on the fibres to avoid further oxidation of the products. This novel method is useful for the synthesis of various terpenes [99].

6.12.2 Electrolytic Reactions Using Diamond Electrodes

Diamond is a solid crystal consisting of sp^3 carbons forming covalent bonds. It has excellent transparency and thermoconductivity as well as mechanical and chemical stability. Although pure diamond is non-conductive, it becomes electrically conductive by doping with an impurity such as boron. Boron-doped diamond films can be used as electrodes, which exhibit excellent and superior electrochemical properties compared to known metal and carbon electrode materials. Accordingly, boron-doped diamond (BDD) electrodes are promising novel electrode materials [100].

6.12.2.1 *Electrochemical Features and Application to Highly Sensitive Electroanalysis*

Wide Potential Window in Aqueous Solution [101,102]: BDD electrodes have a hydrogen overpotential at about −1 V vs. Ag/AgCl, which is comparable to mercury, and an oxygen overpotential at about +2.2 V vs. Ag/AgCl, which is higher than other common anode materials. Its potential window is therefore as wide as 3.2 V and this is a pronounced feature of the BDD electrode. Even in aprotic organic solvents like MeCN, the cathodic window is about 500 mV wider than that of a Pt electrode. Accordingly, such unique properties could result in replacement of either costly noble metals or toxic heavy metals.

Small Background Current (Small Residual Current Density) [101,102]: The capacitance of the BDD electrode surface is a few mF cm^{-2}, which is two orders smaller compared to glassy carbon, and its residual current density is as small as a few hundred nA cm^{-2}. This is attributable to the fact that neither dissolved gas nor ions of supporting salt adsorb on the BDD surface, making redox reactions difficult.

Large Overpotential of Oxygen Reduction [103]: Because of the difficult cathodic reduction of dissolved oxygen, the desired reduction current can be observed even without removal of dissolved oxygen.

Physical and Chemical Stability and Excellent Durability [100,104]: BDD electrodes are basically degradation-free and can be used even in a corrosive solution. Electrolysis can be conducted at high current density (10 A cm^{-2}) and can also be carried out at temperatures up to 600°C. The electrode surface is stable and hardly contaminated by adsorption of impurities.

Fast Electron-Transfer Rate between Redox Species and BDD Electrode [101,104,105]: Outer-sphere one-electron redox systems like $Fe(CN)_6^{4-/3-}$, $IrCl_6^{2-/3-}$ and $Ru(NH_3)_6^{2+/3+}$ show reversible waves in cyclic voltammograms and their electron transfer at the electrode interface is fast.

Unique Electrochemical Selectivity [106]: Oxygen-terminated BDD electrodes prepared by treatment with oxygen plasma or anodic oxidation show selective responses to specific chemical species.

Since BDD electrodes have a wide potential window and extremely small background current in addition to the above features, highly sensitive sensors for detection of biological materials have been developed using BDD electrodes. For example, when an oxygen-terminated BDD electrode is used, a trace amount of dopamine and uric acid can be detected and quantitatively analyzed even in the presence of a large amount of ascorbic acid [106]. Moreover, since the BDD electrode surface is resistant to contamination with adsorbed impurities, it can be used as an electroanalytical detector [107].

Quite recently BDD microelectrodes were developed, which enable *in vivo* pH monitoring for stomach disorder diagnoses as well as *in vivo* assessment of cancerous tumours [108].

6.12.2.2 *Application to Organic Electrosynthesis*

Various anodic substitutions like fluorination, methoxylation, acetoxylation and cyanation of aromatic compounds, heterocycles and sulfides have been achieved so far.

Kolbe electrolysis was also investigated using a BDD anode [109]. Quite recently it was demonstrated that a BDD anode is highly effective for regioselective homocoupling of substituted phenols to provide biphenols, as shown in Eq. 6.30 [110]. BDD electrodes are also superior anodes for the generation of alkoxy radicals from alcohols. Accordingly, anodic 1,2-dimethoxylation of isoeugenol has been realized under microflow conditions [111]. Inagi and co-workers have achieved parallel electrochemical reactions of an alternating copolymer of 9-fluorenol and 9,9-dioctylfluorenone on a BDD bipolar electrode in Et_4NOTs/*i*PrOH, giving a multi-coloured gradient film [112]. This is an excellent application of the wide potential window of BDD electrodes.

BDD anode

$(F_3C)_2CHOH$, $Et_3NMe\ O_3SOMe$

74% 41%

(6.30)

6.12.2.3 Application to Inorganic Electrosynthesis

The BDD electrode is a superior anode for the generation of ozone and hydrogen peroxide from water compared to platinum and iridium oxide anodes [113]. Chlorine gas is readily generated by BDD electrodes and the efficiency for hypochloric acid formation is high, therefore applications in disinfection treatments are expected. The preparation of persulfuric acid from sulfuric acid can be performed using BDD anodes with 70% current efficiency, while the production of ammonia from nitric acid is possible using BDD cathodes [114,115]. Graphite reacts with fluorine gas and graphite fluoride is formed in the bulk of the graphite, resulting in suppression of the Faradaic current. In sharp contrast, only the surface of the BDD electrode is fluorinated, therefore it maintains its electro-conductivity [116]. Since it is stable to hydrogen fluoride, this electrode is expected to be useful for the preparation of fluorine gas.

6.13 PHOTOELECTROLYSIS AND PHOTOCATALYSIS

6.13.1 Photoelectrolysis

When a semiconductor electrode is irradiated, charge separation occurs to generate electron (e)–hole (h^+) pairs. The generated electron works as a reductant, while the hole works as an oxidant. Honda and Fujishima demonstrated for the first time the possibility of such photosensitized electrolysis using the n-type semiconductor TiO_2 and Pt electrodes, as shown in Figure 6.34 [117]. When the surface of the TiO_2 electrode is

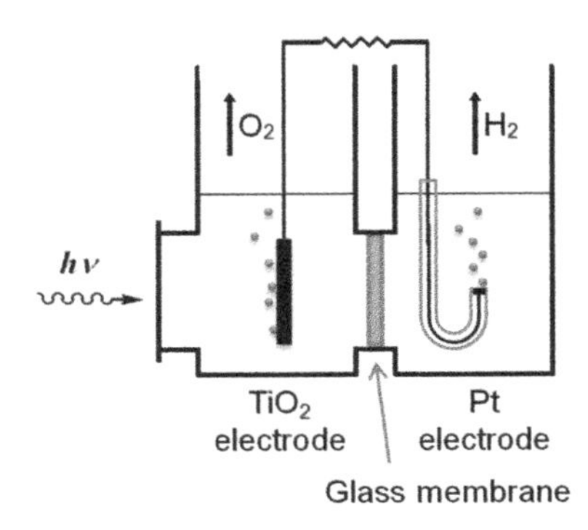

Figure 6.34 Photochemical cell with TiO_2 electrode

irradiated with a wavelength shorter than 410 nm, current flows to generate oxygen and hydrogen at the surfaces of the TiO_2 and Pt electrodes, respectively. TiO_2 is excited under irradiation to generate electron–hole pairs, and the holes in the valence band move to the surface, resulting in oxidation of water, while the electrons in the conduction band move into the bulk and then further move through the external circuit to the counter Pt electrode to reduce protons. This suggests that water is decomposed by visible light into oxygen and hydrogen, without the application of any external voltage, according to Eq. 6.31. This is an innovative breakthrough and is referred to as the Honda–Fujishima effect [117].

$$\begin{aligned} 2h^+ + H_2O &\longrightarrow 1/2\ O_2 + 2H^+ \\ 2e + 2H^+ &\longrightarrow H_2 \\ \hline \text{Total:}\quad H_2O &\xrightarrow{2h\nu} H_2 + 1/2\ O_2 \end{aligned} \tag{6.31}$$

Photoelectrochemical cells have the advantage of producing oxygen and hydrogen in addition to electrical energy. However, there have been few examples of applications of such photoelectrolysis to organic reactions to date.

6.13.2 Photocatalysts

When the circuit between TiO_2 and Pt electrodes is shortened, Pt is eventually deposited on the surface of TiO_2. Accordingly, a particle of TiO_2 with some Pt deposited on it (Figure 6.35) can be considered as a short circuit photoelectrochemical cell. In the case of such a photocatalyst, oxidation and reduction occur simultaneously at two sites on the small particle. Since the oxidation and reduction sites are very close to each other, various unique organic synthetic reactions have been developed [118]. For instance, anodic oxidation of a primary amine generally forms an aldehyde via an imine intermediate, while the same oxidation using a TiO_2 catalyst with Pt deposited on it affords a secondary amine, as shown in Figure 6.35 [119]. In this case, the primary amine is oxidized with the hole generated by photo-excitation to form an aldehyde, which reacts with the unreacted starting amine to form an imine intermediate. The imine is immediately reduced at a Pt site, resulting in the formation of

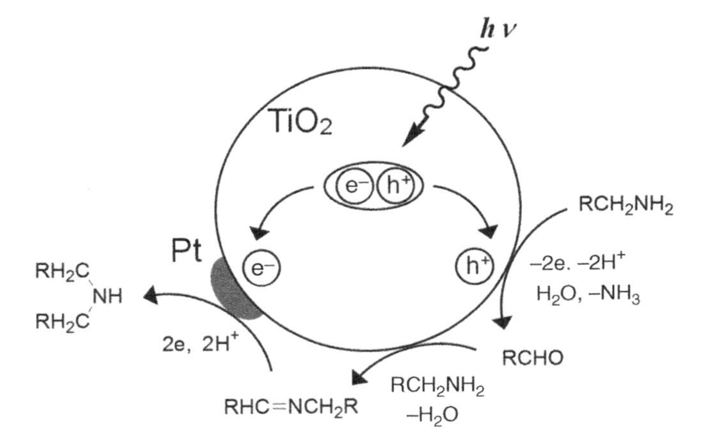

Figure 6.35 Example of photocatalytic reactions

the secondary amine. This is due to both oxidation and reduction occurring at active sites close to each other. By using this principle, the synthesis of optically active pipecolinic acid from optically active lysine was achieved by photocatalytic reaction using platinized semiconductor TiO_2 powder particles, as shown in Eq. 6.32 [120]. Thus, photocatalytic reactions can be considered to be wireless electrolytic reactions.

L-Lysine $H_2N(CH_2)_4CH(NH_2)COOH$ $\xrightarrow[-2H^+]{2h^+}$ $HN{=}CH(CH_2)_3CH(NH_2)COOH$ $\xrightarrow[-NH_3]{H_2O}$ $OHC(CH_2)_3CH(NH_2)COOH$

$\xrightarrow[-H_2O]{}$ (cyclic imine)–COOH $\xrightarrow{2e,\ 2H^+}$ (N–H piperidine)–COOH, L-Pipecolic acid

(6.32)

6.14 ELECTROCHEMICAL POLYMER REACTIONS

As described in Chapter 5, conducting polymers are redox active, especially in the film state, on the electrode surface. In a number of applications, the doping and dedoping process of conducting polymers needs to be reversible to avoid over-oxidation and over-reduction, which may cause undesirable reactions in the polymer structure. Because of this,

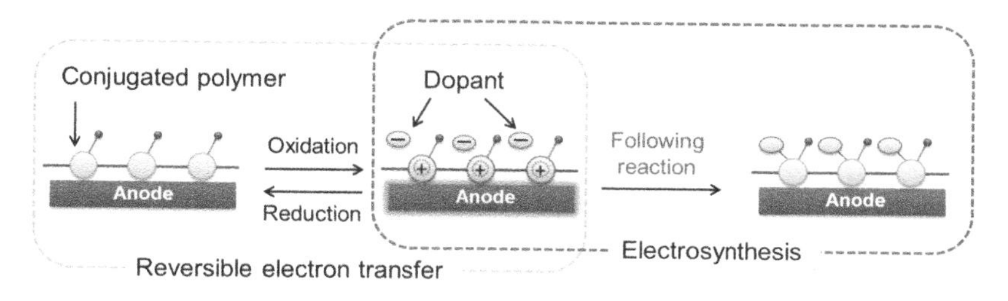

Figure 6.36 Electrochemical polymer reactions

a conducting polymer in its doped state can be regarded as a reactive species capable of undergoing subsequent reactions. The design of specific electrogenerated species generated from specific polymers and the subsequent chemical reactions represents a powerful means of introducing versatile functionalities into polymers. This electrochemical polymer reaction is a kind of post-polymerization functionalization (or simply post-functionalization) (Figure 6.36) [121]. The reaction ratio (or the degree of functionalization) of the repeating reaction sites influences various properties of the resulting conducting polymer, including its absorption, photoluminescence and electrochemistry. As anodic reactions, oxidative halogenations, cyanations and pyridinations of conducting polymers are known to proceed when using the corresponding nucleophilic dopants. Cathodic reaction such as reductive hydrogenation of the 9-fluorenone moiety in a conducting polymer is also possible, so that paired electrochemical polymer reactions in a single cell can be designed and give good current efficiency.

The main advantage of electro-organic synthesis is the ability to rapidly switch the application of electrical potential to the working electrode either on or off. By applying this advantage to electrochemical polymer reactions, the reaction ratio can be finely tuned by precisely controlling the amount of charge passed between the anode and cathode of the electrode. As a result, the physical properties of the conjugated polymer can be tailored by adjusting the reaction ratio.

To progress this solid-phase reaction, the choice of electrolyte is critical. In the electrochemical fluorination of a polyfluorene derivative, ionic liquid hydrogen fluoride salts play the roles of supporting salt, reaction medium and fluorine source, preventing film detachment during electrolysis (Eq. 6.33) [122].

ArS SAr C_8H_{17} C_8H_{17} n

$-2e, -ArSSAr, +2F^-$

$Et_4NF\text{-}5HF$

Ar =

F F C_8H_{17} C_8H_{17} n

(6.33)

REFERENCES

1. Yoshida, J., Kataoka, K., Horcajada, R. and Nagaki, A. (2008) *Chem. Rev.*, **108**, 2265–2299.
2. Ogumi, Z., Nishio, K. and Yoshizawa, Z. (1981) *Electrochim. Acta*, **26**, 1779–1782.
3. (a) Hoormann, D., Kubon, C., Jörissen, J., Kröner, L. and Pütter, H. (2001) *J. Electroanal. Chem.*, **507**, 215–225. (b) Jörissen, J. (2003) *J. Appl. Electrochem.*, **33**, 969–977.
4. (a) Ogumi, Z., Ohashi, S. and Takehara, Z. (1985) *Electrochim. Acta*, **30**, 121–124. (b) Ogumi, Z., Inaba, M., Ohashi, S., Uchida, M. and Takehara, Z. (1988) *Electrochim. Acta*, **33**, 365–369. (c) Ogumi, Z., Inatomi, K., Hinatsu, J.T. and Takehara, Z. (1992) *Electrochim. Acta*, **37**, 1295–1299.
5. (a) Otsuka, K., Shimizu, Y. and Yamanaka, I. (1988) *J. Chem. Soc., Chem. Commun.*, 1272–1273. (b) Otsuka, K., Shimizu, Y. and Yamanaka, I. (1990) *J. Electrochem. Soc.*, **137**, 2076–2081.
6. Tajima, T. and Fuchigami, T. (2005) *J. Am. Chem. Soc.*, **127**, 2848–2849.
7. Tajima, T. and Fuchigami, T. (2005) *Angew. Chem. Int. Ed.*, **44**, 4670–4679.
8. Tajima, T., Kurihara, H. and Fuchigami, T. (2007) *J. Am. Chem. Soc.*, **129**, 6680–6681.
9. Yoshida, J., Nakai, R. and Kawabata, N. (1980) *J. Org.Chem.*, **45**, 5269–5273.
10. Tanaka, H., Kawakami, Y., Goto, K. and Kuroboshi, M. (2001) *Tetrahedron Lett.*, **42**, 445–448.
11. Tanaka, H., Chou, J., Mine, M. and Kuroboshi, K. (2004) *Bull. Chem. Soc. Jpn.*, **77**, 1745–1755.
12. Kubota, J., Ido, T., Kuroboshi, M., Tanaka, H., Uchida, T. and Shimamura, K. (2006) *Tetrahedron*, **62**, 4769–4773.
13. Asami, R., Atobe, M. and Fuchigami, T. (2005) Electrpolymerization of an Immiscible Monomer in Aqueous Electrolytes Using Acoustic Emulsification, *J. Am. Chem. Soc.*, **127**, 13160–13161.
14. Atobe, M., Chen, P.-C. and Nonaka, T. (1998) Ultrasonic Effects on Electroorganic Processes. Part 11. Emulsion and Suspension Electrolyses under Ultrasonic Irradiation, *Electrochemistry*, **66**, 556–559.
15. Chiba, K., Kono, Y., Kim, S., Nishimoto, K. and Tada, M. (2002) A Liquid-phase peptide synthesis in cyclohexane-based biphasic thermomorphic systems, *Chem. Commun.*, **38**, 1766–1767.

16. Yoshida, J. (2005) *Chem. Commun.*, 4509–4516.
17. Yoshida, J., Saito, K., Nokami, T. and Nagaki, A. (2011) *Synlett*, 1189–1194.
18. Malkowsky, I.M., Rommel, C.E., Fröhlich, R., Griesbach, U., Pütter, H. and Waldvogel, S.R. (2006) *Chem. Eur. J.*, **12**, 7482–7488.
19. Hudson, C.M. and Moeller, K.D. (1994) *J. Am. Chem. Soc.*, **116**, 3347–3356.
20. Vögtle, F. and Stoddart, J.F. (2000) *Supercritical Fluids for Organic Synthesis*, Wiley-VCH Verlag GmbH, Weinheim.
21. Jessop, P.G. and Leitner, W. (1999) *Chemical Synthesis Using Supercritical Fluids*, Wiley-VCH Verlag GmbH, Weinheim.
22. Kendall, J.L., Canelas, D.A., Young, J.L. and DeSimone, J.M. (1999) Polymerizations in Supercritical Carbon Dioxide, *Chem. Rev.*, **99**, 543–564.
23. Jun, J. and Fedkiw, P.S. (2001) Ionic conductivity of alkali-metal salts in sub- and supercritical carbon dioxide plus methanol mixtures, *J. Electroanal. Chem.*, **515**, 113–122.
24. Wu, W., Zhang, J., Han, B., Chen, J., Liu, Z., Jiang, T., He, J. and Li, W. (2003) Solubility of room-temperature ionic liquid in supercritical CO_2 with and without organic compounds, *Chem. Commun.*, **39**, 1412–1413.
25. Yan, H., Sato, T., Komago, D., Yamaguchi, A., Oyaizu, K., Yuasa, M. and Otake, K. (2005) Electrochemical Synthesis of a Polypyrrole Thin Film Using Supercritical Carbon Dioxide as a Solvent, *Langmuir*, **21**, 12303–12308.
26. Sasaki, A., Kudoh, H., Senboku, H. and Tokuda, M. (1998) *Electrochemical Carboxylation of Several Organic Halides in Supercritical Carbon Dioxide, Novel Trends in Electroorganic Synthesis* (ed. S. Torii), Springer-Verlag, Tokyo, pp. 245–246.
27. Dombro Jr, R.A., Prentice, G.A. and McHugh, M.A. (1988) Electro-Organic Synthesis in Supercritical Organic Mixtures, *J. Electrochem. Soc.*, **135**, 2219–2223.
28. Mori, T., Li, M., Kobayashi, A. and Okahata, Y. (2002) Reversible Control of Enzymatic Transglycosylations in Supercritical Fluoroform using a Lipid-Coated b-D-Galactosidase, *J. Am. Chem. Soc.*, **124**, 1188–1189.
29. Atobe, M., Ohsuka, H. and Fuchigami, T. (2004) Electrochemical Synthesis of Polypyrrole and Polythiophene in Supercritical Trifluoromethane, *Chem. Lett.*, **33**, 618–619.
30. Atobe, M., Iizuka, S., Fuchigami, T. and Yamamoto, H. (2007) Preparation of Nanostructured Conjugated Polymers Using Template Electrochemical Polymerization in Supercritical Fluids, *Chem. Lett.*, **36**, 1448–1449.
31. (a) Ohno, H. C (ed.) (2005) *Electrochemical Aspects of Ionic Liquids*, Wiley. (b) Ohno, H. (ed.) (2011) Electrochemical Aspects of Ionic Liquids, 2nd edn, Wiley.
32. Welton, T. (1999) *Chem. Rev.* **99**, 2071–2084.
33. Wassersheid, P. and Keim, W. (2000) *Angew Chem Int. Ed.*, **39**, 3772–3789.
34. Rogers, R.D., Seddon, K.R. and Volkov, S. (eds) (2002) *Green Industrial Applications of Ionic Liquids*, Kluwer Academic.
35. Hapiot, P. and Lagrost, C. (2008) *Chem. Rev.*, **108**, 2238–2264.
36. Sweeny, B.K. and Peters, D.G. (2001) *Electrochem. Commun.*, **3**, 712–715.
37. Doherty, A.P. and Brooks, C.A. (2004) *Electrochim Acta*, **49**, 3821–3826.
38. Lagrost, C., Hapiot, P. and Vaultier, M. (2005) *Green Chem.*, **7**, 468–474.
39. Mellah, M., Gmouh, S., Vaultie, M. and Jouikov, V. (2003) *Electrochem. Commun.*, **5**, 591–593.

40. Pachon, P.L., Elsevier, C.J. and Rothenberg, G. (2006) *Adv. Synth. Catal.*, **348**, 1705–1710.
41. She, Y., Atobe, M., Tajim, T. and Fuchigami, T. (2004) *Electrochemistry*, **72**, 849–851.
42. Yang, H., Gu, Y., Deng, Y. and Shi, F. (2002) *Chem. Commun.*, 274–275.
43. Feroci, M., Orsini, M., Rossi, L., Sotgiu, G. and Inesi, A. (2007) *J. Org. Chem.*, **72**, 200–203.
44. Mellah, M., Zeitouny, J., Gmouh, S., Vaultier, M. and Jouikov, V. (2005) *Electrochem. Commun.*, **7**, 869–874.
45. Chowdhury, S., Mohan, R.S. and Scott, J.L. (2007) *Tetrahedron*, **63**, 2363–2389.
46. Feroci, M., Chiarotto, M., Orsini, I. Sotgiu, M. and Inesi, G.A. (2008) *Adv. Synth. Catal.*, **350**, 1355–1359.
47. Feroci, M., Elinson, M.N., Rossi, L. and Inesi, A. (2009) *Electrochem Commun.*, **11**, 1523–1526.
48. Villagrán, C., Banks, C.E., Pitner, W.R., Hardacre, C. and Compton, R.G. (2005) *Ultrason. Sonochem.*, **12**, 423–428.
49. (a) Fuchigam, T. and Inagi, S. (2011) *Chem. Commun.*, **47**, 10211–10223. (b) Fuchigami, T. (2000) Electrochemical Partial Fluorination, in *Organic Electrochemistry*, 4th edn (eds H. Lund and O. Hammerich), Marcel Dekker, New York, Chapter 25.
50. (a) Momota, K. (1999) *Yoyuen (Molten Salts)*, **39**, 7–22. (b) Momota, K., Morita, M. and Matsuda, Y. (1993) *Electrochim. Acta*, **38**, 1123–1130.
51. Fuchigami, T. (2007) *J. Fluorine Chem.*, **128**, 311–316.
52. Meurs, H.H. and Eilenberg, W. (1991) *Tetrahedron*, **47**, 705.
53. (a) Momota, K., Horio, H., Kato, K., Morita, M. and Matsuda, Y. (1995) *Electrochim. Acta*, **40**, 233–240. (b) Momota, K., Yonezawa, T., Hayakawa, Y., Kato, K., Morita, M. and Matsuda, Y. (1995) *J. Appl. Electrochem.*, **25**, 651–658.
54. (a) Momota, K., Mukai, K., Kato, K. and Morita, M. (1998) *Electrochim. Acta*, **43**, 2503–2514. (b) Momota, K., Mukai, K., Kato, K. and Morita, M. (1998) *J. Fluorine Chem.*, **87**, 173–178.
55. Yoneda, N., Chen, S.-Q., Hatakeyama, T., Hara, S. and Fukuhara, T. (1994) *Chem. Lett.*, 849–850.
56. Chen, S.-Q., Hatakeyama, T., Fukuhara, T., Hara, S. and Yoneda, N. (1997) *Electrochim. Acta*, **42**, 1951–1960.
57. Hara, S., Chen, S.-Q., Hoshio, T., Fukuhara, T. and Yoneda, N. (1996) *Tetrahedron Lett.*, **37**, 8511–8514.
58. Fukuhara, T., Akiyama, Y., Yoneda, N., Tada, T. and Hara, S. (2002) *Terahedron Lett.*, **43**, 6583–6585.
59. Aoyama, M., Fukuhara, T. and Hara, S. (2008) *J. Org. Chem.*, **73**, 4186–4189.
60. (a) Hasegawa, M., Ishii, H. and Fuchigami, T. (2002) *Tetrahedron Lett.*, **43**, 1503–1505. (b) Hasegawa, M., Ishii, H., Cao, Y. and Fuchigami, T. (2006) *J. Electrochem. Soc.*, **153**, D162–D166.
61. Hasegawa, M., Ishii, H. and Fuchigami, T. (2003) *Green Chem.*, **5**, 512–515.
62. Sunaga, T., Atobe, M., Inagi, S. and Fuchigami, T. (2009) *Chem. Commun.*, 956–958.
63. (a) Hou, Y. and Fuchigami, T. (2000) *J. Electrochem. Soc.*, **147**, 4567–4572. (b) Shaaban, M.R., Ishii, H. and Fuchigami, T. (2000) *J. Org. Chem.*, **65**, 8685–8689. (c) Dawood, K.M. and Fuchigami, T. (1999) *J. Org. Chem.*, **64**, 138–143.

64. Sawamura, T., Inagi, S. and Fuchigami, T. (2009) *J. Electrochem. Soc.*, **156**, E26–E28.
65. Sawamura, T., Kuribayashi, S., Inagi, S. and Fuchigami, T. (2010) *Org. Lett.*, **12**, 644–646.
66. Sawamura, T., Kuribayashi, S., Inagi, S. and Fuchigami, T. (2010) *Adv. Synth. Catal.*, **352**, 2757–2760.
67. Lu, W., Fadeev, A.G., Qi, B., Smela, E., Mattes, B.R., Ding, J., Spinks, G.M., Mazurkiewicz, J., Zhou, D., Wallace, G.G., MacFarlane, D.R., Forsyth, S.A. and Forsyth, M. (2002) *Science*, **297**, 983–987.
68. Sekiguchi, K., Atobe, M. and Fuchigami, T. (2002) *Electrochem. Commun.*, **4**, 881–885.
69. Sekiguchi, K., Atobe, M. and Fuchigami, T. (2003) *J. Electroanal. Chem.*, **557**, 1–7.
70. Naudin, E., Ho, H.A., Branchaud, S., Breau, L. and Belanger, D. (2002) *J. Phys. Chem. B*, **106**, 10585–10593.
71. Randriamahazaka, H., Plesse, C. and Chevrot, D. (2004) *Electrochem. Commun.*, **6**, 299–305.
72. Abedin, S.Z.E., Borissenko, N. and Endres, F. (2004) *Electrochem. Commun.*, **6**, 422–426.
73. Juttner, K. (2007) Technical Scale of Electrochemistry, in *Electrochemical Engineering* (eds D.D. Macdonald and P. Schmuki), Wiley-VCH Verlag GmbH, Chapter 1.
74. Paddon, C.A., Pritchard, G.J., Thiemann, T. and Marken, F. (2002) Paired electrosyn-thesis: micro-flow cell processes with and without added electrolyte, *Electrochem. Commun.*, **4**, 825–831.
75. Horii, D., Atobe, M., Fuchigami, T. and Marken, F. (2006) Self-supported Methoxylation and Acetoxylation Electrosynthesis Using a Simple Thin-layer Flow Cell, *J. Electrochem. Soc.*, **153**, D143–D147.
76. Suga, S., Okajima, M., Fujiwara, K. and Yoshida, J. (2001) A New Approach to Conventional and Combinatorial Organic Syntheses Using Electrochemical Micro Flow Systems, *J. Am. Chem. Soc.*, **123**, 7941–7942.
77. Horii, D., Fuchigami, T. and Atobe, M. (2007) A New Approach to Anodic Substitution Reaction Using Parallel Laminar Flow in a Micro-Flow Reactor, *J. Am. Chem. Soc.*, **129**, 11692–11693.
78. Amemiya, F., Fuse, K., Fuchigami, T. and Atobe, M. (2010) Chemoselective Reaction System Using a Two Inlets Micro-flow Reactor: Application to Reductive Carbonyl Allylation, *Chem. Commun.*, **46**, 2730–2732.
79. Walton, D.J. and Phull, S.S. (1996) Sonoelectrochemistry, in *Advances in Sonochemistry*, Vol. 4 (ed. T.J. Mason), JAI Press, London.
80. Lorimer, P. and Mason, T.J. (1999) The application of Ultrasound in Electroplating, *Electrochemistry*, **67**, 924–929.
81. Suslick, K.S., Gawienowski, J.J., Schubert, P.F. and Wang, H.H. (1983) Alkane sonochemistry, *J. Phys. Chem.*, **87**, 2299.
82. Matsuda, K., Atobe, M. and Nonaka, T. (1994) Ultrasonic Effects on Electroorganic Processes. Part1. Product-selectivity in Electroreduction of Benzaldehydes, *Chem. Lett.*, **23**, 1619–1622.
83. Atobe, M., Matsuda, K. and Nonaka, T. (1996) Ultrasonic Effects on Electroorganic Processes. Part 4. Theoretical and Experimental Studies on Product-selectivity in Electroreduction of Benzaldehyde and Benzoic Acid, *Electroanalysis*, **8**, 784–788.
84. Atobe, M. and Nonaka, T. (1997) *Chem. Lett.*, **26**, 323–324.

85. Heinze, J. (2001) Electrochemistry of Conducting Polymers, in *Organic Electrochemistry* (eds H. Lund and O. Hammerich), Marcel Dekker, New York.
86. Osawa, S., Ito, M., Tanaka, K. and Kuwano, J. (1987) *Synth. Met.*, **18**, 145–150.
87. Atobe, M. and Nonaka, T. (1998) New Developments in Sonoelectrochemistry, *Nippon Kagaku Kaishi*, 219–230.
88. Atobe, M., Kaburagi, T. and Nonaka, T. (1999) Ultrasonic Effects on Electroorganic Processes. Part 13. A Role of Ultrasonic Cavitation in Electrooxidative Polymerization of Aniline, *Electrochemistry*, **67**, 1114–1116.
89. Atobe, M., Chowdhury, A.N., Fuchigami, T. and Nonaka, T. (2003) Preparation of Conducting Polyaniline Colloids under Ultrasonication, *Ultrason. Sonochem.*, **10**, 77–80.
90. Chowdhury, A.N., Atobe, M. and Nonaka, T. (2004) Studies on Solution and Solution-Cast Film of Polyaniline Colloids Prepared in the Absence and Presence of Ultrasonic Irradiation, *Ultrason. Sonochem.*, **11**, 77–82.
91. Park, J.E., Atobe, M. and Fuchigami, T. (2005) Sonochemical Synthesis of Conducting Polymer-Metal Nanoparticles Nanocomposite, *Electrochim. Acta*, **51**, 849–854.
92. Park, J.E., Atobe, M. and Fuchigami, T. (2005) Sonochemical Synthesis of Inorganic-organic Hybride Nanocomposite Based on Gold Nanoparticles and Polypyrrole, *Chem. Lett.*, **34**, 96–97.
93. Atobe, M., Tsuji, H., Asami, R. and Fuchigami, T. (2006) A Study on Doping-undoping Properties of Polypyrrole Films Electropolymerized under Ultrasonication, *J. Electrochem. Soc.*, **153**, D10–D13.
94. Taouil, A.E., Lallemand, F., Hihn, J.Y. and Blondeau-Patissier, V. (2011) Electrosynthesis and characterization of conducting polypyrrole elaborated under high frequency ultrasound irradiation, *Ultrason. Sonochem.*, **18**, 907–910.
95. Taouil, A.E., Lallemand, F., Hihn, J.Y., Melot, J.M., Blondeau-Patissier, V. and Lakard, B. (2011) Doping properties of PEDOT films electrosynthesized under high frequency ultrasound irradiation, *Ultrason. Sonochem.*, **18**, 140–148.
96. (a) Kunugi, Y., Fuchigami, T., Tien, H.-J. and Nonaka, T. (1989) *Chem. Lett.*, 757–760. (b) Kunugi, Y., Fuchigami, T. and Nonaka, T. (1990) *J. Electroanal. Chem.*, **287**, 385–388.
97. Kunugi, Y., Nonaka, T., Chong, Y.-B. and Watanabe, N. (1992) *Electrochim. Acta*, **37**, 353–355.
98. Chiba, K., Jinno, M., Kuramoto, R. and Tada, M. (1998) *Tetrahedron Lett.*, **39**, 5527–5539.
99. Chiba, K., Fukuda, M., Kim, S., Kitano, Y. and Tada, M. (1999) *J. Org. Chem.*, **64**, 7654–7656.
100. (a) Brillas, E. and Martínez-Huitle, C.A. (eds) (2011) *Synthetic Diamond Films – Preparation, Electrochemistry, Characterization and Applications*, Wiley-VCH Verlag GmbH, Weinheim. (b) Fujishima, A. and Einaga, Y. (eds) (2005) *Diamond Electrochemistry*, Elsevier and BKC.
101. (a) Bouamrane, F., Tadjeddine, A., Butler, J.E., Tenne, R. and Lévy-Clément, C. (1996) *J. Electroanal. Chem.*, **405**, 95–99. (b) Chen, Q., Granger, M.C., Lister, T.E. and Swain, G.M. (1997) *J. Electrochem. Soc.*, **144**, 3806–3812.
102. Martin, H.B., Argoitia, A., Landau, U., Anderson, A.B. and Angus, J.C. (1996) *J. Electrochem. Soc.*, **143**, L133–L136.

103. Yano, Y., Tryk, D.A., Hashimoto, K. and Fujishima, A. (1998) *J. Electrochem. Soc.*, **145**, 1870–1876.
104. Swain, G.M. (1994) *J. Electrochem. Soc.*, **141**, 3382–3393.
105. DeClemnts, R. and Swain, G.M. (1997) *J. Electrochem. Soc.*, **144**, 856–866.
106. Popa, E., Notsu, H., Miwa, T., Tryk, D.A. and Fujishima, A. (1999) *Electrochem. Solids. Lett.*, **2**, 49–51.
107. Jolly, S., Koppang, M., Jackson, T. and Swain, G.M. (1997) *Anal. Chem.*, **69**, 4099–4107.
108. Fierrol S., Yoshikawa, M., Nagano, O., Yoshimi, K., Saya, H. and Einaga, Y. (2012) *Sci. Rep.*, **2**, 901.
109. Wadhawan, J.D., Wadhawan, J.D., Campo, F.J., Compton, R.G., Foord, J.S., Marken, F., Bull, S.D., Davies, S.G., Walton, D.J. and Ryley, S. (2001) *J. Electroanal. Chem.*, **507**, 135–143.
110. (a) Kirste, A., Nieger, M., Malkowsky, I.M., Stecker, F., Fischer, A. and Waldvogel, S.R. (2009) *Chem. Eur. J.*, **15**, 2273–2277. (b) Kirste, A., Schnakenburg, G. and Waldvogel, S.R. (2011) *Org. Lett.*, **13**, 3126–3129. (c) Kirste, A., Elsler, B., Schnakenburg, G. and Waldvogel, S.R. (2012) *J. Am. Chem. Soc.*, **134**, 3571–3576.
111. Sumi, T., Saito, T., Natsui, K., Yamamoto, T., Atobe, M., Einaga, Y. and Nishiyama, S. (2012) *Angew. Chem. Int. Ed.*, **51**, 5443–5446.
112. Inagi, S., Nagai, H., Tomita, I. and Fuchigami, T. (2013) *Angew. Chem. Int. Ed.*, **52**, 6616–6619.
113. Katsuki, N., Takahashi, E., Toyota, M., Kurosu, T., Lida, M., Wakita, S., Nishiki, Y. and Shimamune, T. (1998) *J. Electrochem. Soc.*, **145**, 2358–2362.
114. Ferro, S., De Battisti, A., Comninellis, Ch. and Haenni, W. (2000) *J. Electrochem. Soc.*, **147**, 2614–2619.
115. (a) Reuben, C., Galun, E., Cohen, H., Tenne, R., Kalish, R., Muraki, Y., Hashimoto, K., Fujishima, A., Butler, J.M. and Lévy-Clément, C. (1995) *J. Electroanal. Chem.*, **396**, 233–239. (b) Tenne, R., Patel, K., Hashimoto, K. and Fujishima, A. (1993) *J. Electroanal. Chem.*, **347**, 409–415.
116. Ando, T., Yamamoto, K., Kamo, M., Sato, Y., Takamatsu, Y., Kawasaki, S., Okino, F. and Touhara, H. (1995) *J. Chem. Soc., Faraday Trans.*, **91**, 3209–3212.
117. Fujishima, A. and Honda, K. (1972) *Nature*, **238**, 37–38.
118. Fox, M.N. (1983) *Acc. Chem. Res.*, **16**, 314–332.
119. Nishimoto, S., Ohtani, B., Yoshikawa, T. and Kagiya, T. (1983) *J. Am. Chem. Soc.*, **105**, 7180–7182.
120. Ohtani, B., Tsuru, S., Nishimoto, S. and Kagiya, T. (1990) *J. Org. Chem.*, **55**, 5551–5553.
121. Inagi, S. and Fuchigami, T. (2014) *Macromol. Rapid Commun.*, **35**, 854–867.
122. Inagi, S., Hayashi, S. and Fuchigami, T. (2009) *Chem. Commun.*, 1718–1720.

7

Related Fields of Organic Electrochemistry

Shinsuke Inagi and Toshio Fuchigami

The world is facing severe problems such as environmental problems, exhaustion of resources and energy-relevant problems. Organic electrochemistry, including electrosynthesis, has the potential to solve these problems. This chapter describes the contribution of organic electrochemistry to organic electronics, the reuse of biomass, C1-chemistry and environmental clean-up.

7.1 APPLICATION IN ORGANIC ELECTRONIC DEVICES

Organic molecules and polymers with unique electronic properties have many device applications in luminescent materials, conductive materials and energy storage. These organic-based materials are superior to inorganic materials in terms of lightness, flexibility and cost. Furthermore, the millions of molecular designs for organic materials make it easy to

Fundamentals and Applications of Organic Electrochemistry: Synthesis, Materials, Devices, First Edition. Toshio Fuchigami, Mahito Atobe and Shinsuke Inagi.

tailor their optoelectronic properties. The basic principles of organic electronics devices are described below from the point of view of organic electrochemistry.

7.1.1 Organic Electroluminescence [1,2]

Organic photoluminescent materials generally emit through the relaxation process of once photoexcited electrons to their ground state (Figure 7.1). The excited state can also be formed electrochemically. When enough voltage is applied to the anode and cathode sandwiching the organic emitting layer (thin film), an injection of holes at the anode and an injection of electrons at the cathode occur. The charges migrate inside the layer and recombine to emit light (Figure 7.2). In order to promote smooth electron injection from the electrode surface to the LUMO of the emission layer, an additional thin layer, i.e. the electron-transport layer, is introduced between the interfaces. In a similar manner, the hole-transport layer is used for hole injection from the electrode to the HOMO of the emission layer. Although these transport layers are important for injecting charge carriers into the emission layer, they should not interrupt the emission of the emission layer. Each organic layer is formed by chemical vapour desorption (CVD) of crystalline molecules or a wet-process of polymer solution.

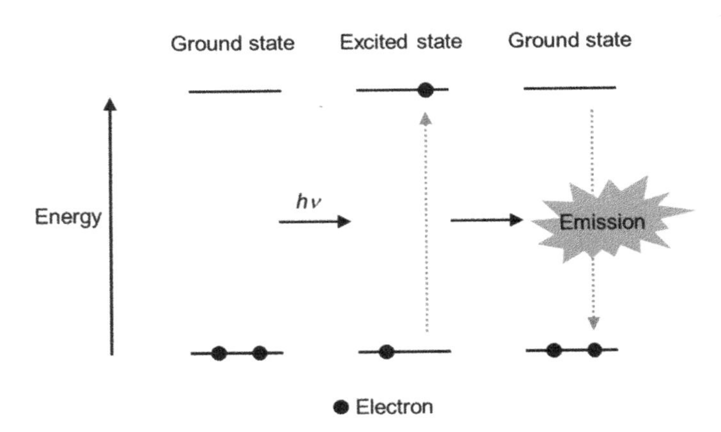

Figure 7.1 Mechanism of photoluminescence

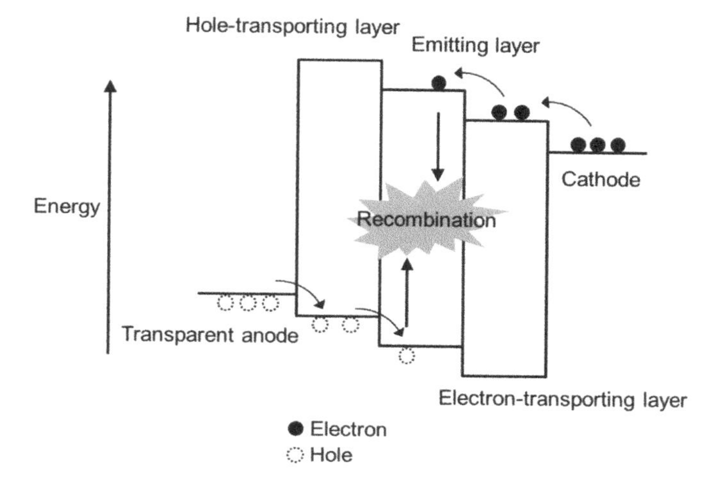

Figure 7.2 Mechanism of electroluminescence

7.1.2 Organic Photovoltaic Cells [3,4]

The principle of photoelectric conversion in an organic photovoltaic cell (OPVC) is the inverse process of organic electroluminescence (OEL). OPVC generates electric energy through the charge separation state obtained by photo-excitation, whereas electric energy causes electron injection and hole injection, and the subsequent charge recombination results in emission in the OEL process.

To induce a charge separation state, electron-donating and electron-accepting layers are necessary in the thin film of the OPVC (Figure 7.3). When the photo-excited state (exciton) is generated in the donor layer by solar energy at the near interface of the donor layer and the acceptor layer, the excited electron may migrate spontaneously to the LUMO of the acceptor layer, resulting in the formation of a charge-separated state. The charge carriers then migrate by hopping through each layer to the electrode terminals and this process drives the photovoltaic cell.

The design of the junction of the layers is important to induce charge separation efficiently before deactivation of the exciton. The construction of a path to the electrode surface of the generated charge carriers is also important for high photo-electron conversion efficiency.

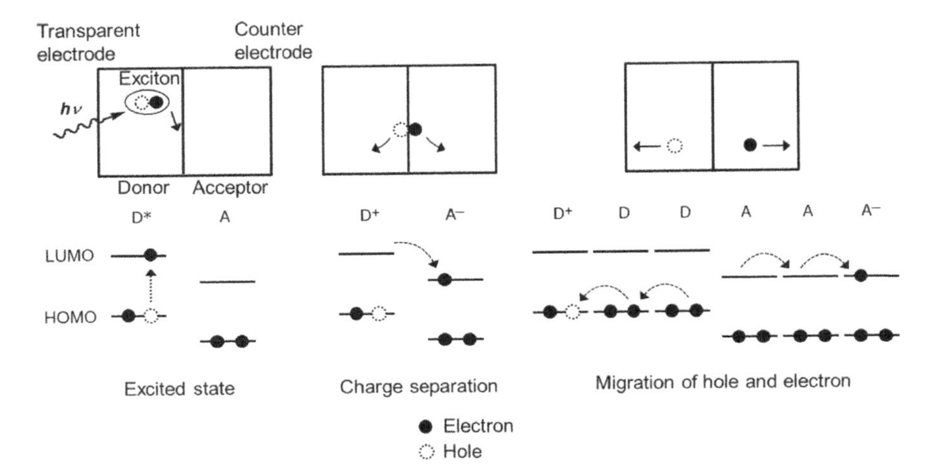

Figure 7.3 Photo-electronic conversion process

7.1.3 Dye-sensitized Solar Cells [5,6]

Semi-conductor electrode like metal oxide can generate electric energy based on the principle of photocatalysts. A photo-excited electron in a semi-conductor electrode reaches the electric circuit and the remaining hole receives electrons from a redox system in the electrolyte, followed by electron transfer from the counter electrode to the redox system. Because of the relatively wide band gap of semi-conductor electrodes, UV light is usually employed to drive this type of photovoltaic cell. Dye-sensitized solar cells (DSSCs) work in the visible region of solar light using organic dye on a semi-conductor electrode as the photo-sensitizer. The organic dye should have a higher LUMO level than the semi-conductor electrode and a lower HOMO level than the redox system in the electrolyte. A typical DSSC system, known as the Grätzel cell, is composed of a transparent anode, titanium oxide (TiO_2) coated with ruthenium complex, an I^-/I_3^- redox system in electrolyte and a Pt cathode (Figure 7.4). The voltage generated in the DSSC corresponds to the difference in energy levels between the Fermi-level of TiO_2 and the standard electrode potential of the I^-/I_3^- redox system.

7.1.4 Organic Transistors [7,8]

Organic transistors, in which organic material is used for the semi-conductor layer, have been widely explored. A model of an organic field

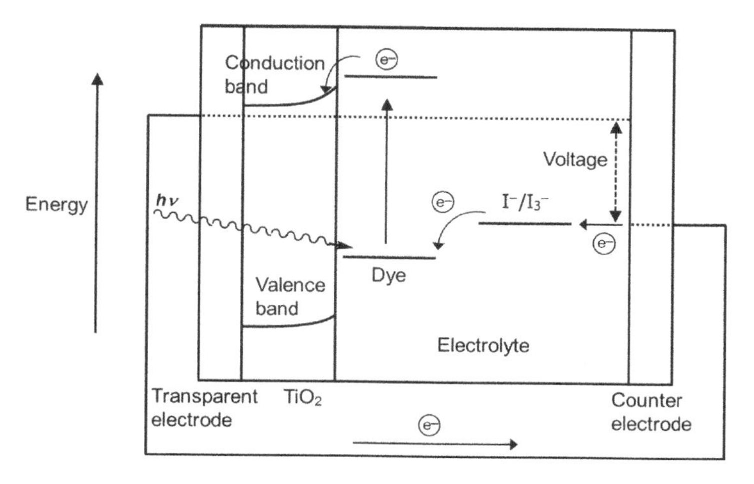

Figure 7.4 Structure of Grätzel cell

effect transistor (OFET) is shown in Figure 7.5. The mechanism of the OFET is as follows: a gate voltage induces charge carriers at the interface of the organic and insulating layers, resulting in decreasing resistance between the source electrode and the drain electrode.

Polycyclic aromatic hydrocarbons and other conjugated molecules are suitable for this purpose. Intermolecular carrier mobility through the organic layer is an important factor, therefore orientation control of the molecules and the design of the junction between the organic layer and the electrodes are important. A number of polymer-based materials (conjugated polymers) for OFET have been developed. A wet-process is available and is useful for fabricating large area devices at low cost, although carrier mobility is generally lower than for highly oriented small molecules.

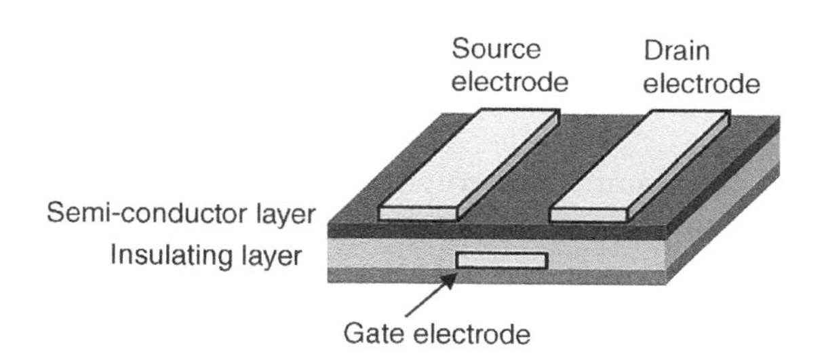

Figure 7.5 Illustration of top-contact OFET

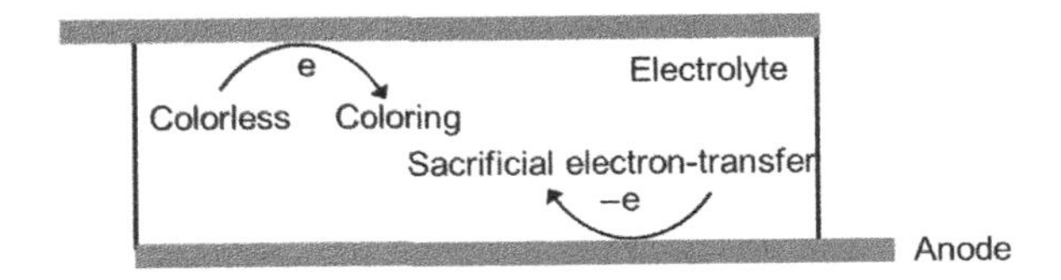

Figure 7.6 Mechanism of electrochromic device

7.1.5 Electrochromic Devices [9]

Electrochemical reaction of organic molecules sometimes accompanies drastic colour change. Reversible redox behaviour with colour change is applicable to electrochromic devices. This colour developing and reducing property, which is different to luminescence phenomena, is suitable for electronic paper applications. Once coloured by the electrochemical reaction, the given image and information remain even after a power cut.

Figure 7.6 shows a typical electrochromic device consisting of a sandwiched structure of electrolyte containing a chromic compound with anode and cathode plates. To visualize colour changes, a transparent electrode should be used for the electrochemical reaction. The chromic compounds used need to have reversible redox properties, i.e. their radical anion or radical cation state should be stable. They may be fixed on a transparent electrode to avoid colour reducing once a coloured state is generated at a counter electrode. As described in section 5.8.6, the use of a stable doping state and the reversible colour changes of conducting polymers for electrochromic applications is very convenient. In addition, film-forming property of electropolymerized conducting polymers is advantageous. Multicoloured electrochromic devices are obtained by stacking each cell, showing their individual colours.

7.1.6 Conducting Polymer-based Capacitors

Conventional aluminium electrolytic capacitors are composed of an aluminium anode, aluminium oxide film, electrolytic solution and an aluminium cathode (Figure 7.7). Aluminium solid capacitors replace the electrolytic solution with solid conductive materials such as β-MnO_2, organic conductors and conducting polymers. In contrast to the ionic-conducting mechanism of electrolytic solutions, solid capacitors are

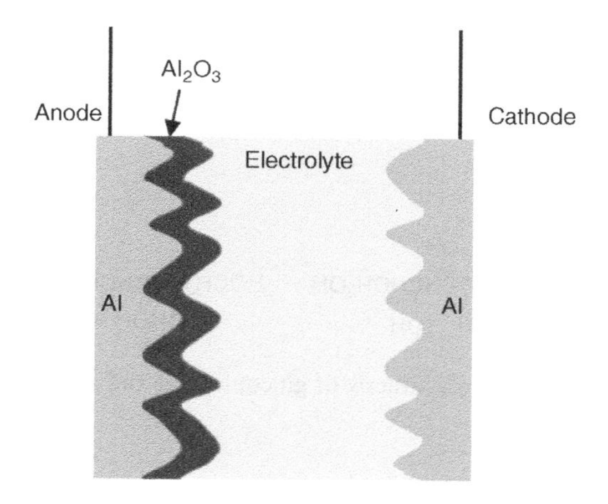

Figure 7.7 Aluminium electrolytic capacitor

driven by an electron-conducting mechanism, and thus conductivity is greatly improved. Furthermore, the thermal properties of solid capacitors are superior to those of conventional ones.

7.2 ELECTROCHEMICAL CONVERSION OF BIOMASS TO VALUABLE MATERIALS

Biomass is a renewable organic resource and therefore it does not cause a depletion problem, unlike fossil resources such as petroleum and coal. In the 21st century, the social demands of the effective utilization of biomass, which can be transformed to high-performance chemicals and fuels, have rapidly increased. Chemical transformation of biomass has to be performed using energy conservation technology and with low emissions, and accordingly various trials to achieve such goals have been carried out using catalytic and enzymatic reactions. As mentioned earlier, electrochemical reactions have the advantage of both energy conservation and low emissions. Intensive study of the electrochemical processes applied to the chemical transformation of biomass started more than 30 years ago, and the transformation of biomass to high-performance chemicals was investigated as a national project in the USA [10]. Today, the study of biomass transformation is increasingly important.

The main components of plant-based biomass are cellulose and lignin, which are polymers. Since neither of these are suitable for electrolysis

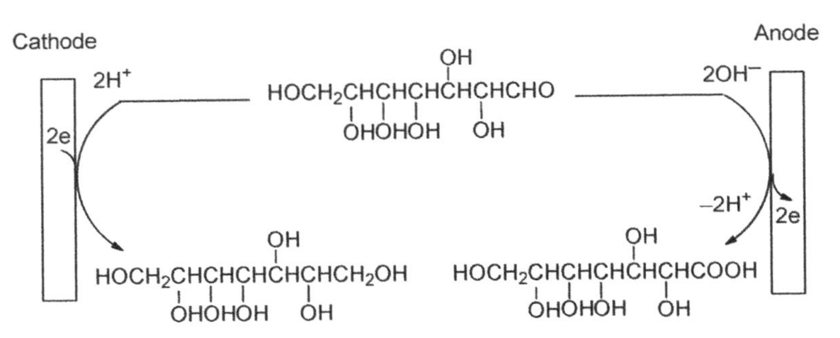

Figure 7.8 Paired electrosynthesis of gluconic acid and sorbitol from glucose

because of their insolubility in solvents like water, biomass transformation to useful chemicals has been attempted through either hydrolysis followed by electrolysis or indirect electrolysis with appropriate mediators.

Since water-soluble glucose is available by hydrolysis of cellulose, its transformation to gluconic acid and sorbitol has been achieved with 90% and 50% current efficiency, respectively, by electrochemical oxidation and reduction of glucose using an undivided cell, as shown in Figure 7.8. Each of these electrochemical processes has been practiced separately on a commercial scale.

Baizer and his co-workers realized the paired electrosynthesis of gluconic acid and sorbitol from glucose using both electrode reactions [11].

Synthesis of dialdehyde starch was achieved by oxidation of starch with IO_3^- mediator using the Ex-cell indirect electrolysis method (100% current efficiency). Other various electrochemical transformation processes of biomass have also been developed, for example paired electrosynthesis of furfuryl alcohol and 2-furancarboxylic acid from furfural derived from hemicellulose, and electrosynthesis of acetone, 2-butene and methylethyl ketone from 2,3-butanediol, which is readily available through fermentation of glucose and xylose [12].

Furthermore, anodic oxidative conversion of lignin to small molecules has been attempted, and aromatic compounds like vanillin obtained [13]. Waldvogel and BASF developed a process for the electrochemical degradation of lignin at a boron-doped diamond (BDD) anode in an aqueous solution to produce hydroxybenzaldehyde derivatives like vanillin and/or phenol derivatives in higher than 5% yield by weight [14]. Quite recently, oxidation of lignosulfonic acid at silver or nickel anode in an aqueous alkaline solution using a flow-type cell has been investigated in order to increase the selectivity and yields of useful compounds like vanillin, vanillic acid and guaiacol (Schmitt and Waldvogel, unpublished).

Efficient use of biomass should be increasingly important in the world, and electrolysis is expected to be a powerful tool to achieve this beneficial utilization of natural resources.

7.3 APPLICATION TO C1 CHEMISTRY

The concept of C1 chemistry is the synthesis of various useful organic compounds through the formation of carbon–carbon bonds using C1 compounds as a starting substrate or by the introduction of various atoms to C1 compounds. In the 1980s, related studies were carried out as a national project, and recently C1 chemistry has again attracted much attention in relation to green sustainable chemistry.

A typical example of C1 chemistry, the fixation of CO_2, is a highly important subject. Since CO_2 is a final product of the combustion processes, the reverse conversion and use of CO_2 are highly important research subjects. However, the reduction potential of CO_2 is very negative, and hence the cathodic evolution of hydrogen takes place as a competitive reaction, reducing the current efficiency and product yield. The cathodic reduction has therefore been intensively studied using various cathode materials. Among these materials, copper was found to be most effective for the reduction of CO_2 [15]. In order to increase efficiency, cathodic reduction of CO_2 was carried out under high pressure. Cathodic reduction of CO_2 provides formic acid, oxalic acid, methane and so on (Eq. 7.1).

$$CO_2 \xrightarrow{e} \begin{matrix} COOH \\ | \\ COOH \end{matrix} + HCOOH + CH_4 \qquad (7.1)$$

Quite recently, Nakata, Einaga and co-workers achieved the electrochemical reduction of CO_2 in seawater using a BDD electrode under ambient conditions to provide formaldehyde selectively with high current efficiency (74%), as shown in Eq. 7.2 [16]. The high current efficiency is attributable to the wide potential window and the sp^3-bonded carbon of the BDD.

$$CO_2 \xrightarrow[\text{seawater, BDD cathode}]{4e,\ 4H^+} \underset{\text{74\% current efficiency}}{HCHO} \qquad (7.2)$$

Cathodic reduction of CO has also been carried out to form squaric (quadratic) acid, as shown in Eq. 7.3 [17].

$$\text{CO} \xrightarrow{e} \text{(squaric acid: cyclobutene ring with two C=O and two OH)} \qquad (7.3)$$

7.4 ENVIRONMENTAL CLEANUP

Electrochemical treatment involving anodic decomposition is a highly promising method for the reduction of toxic pollutants dissolved in waste water. It is important to select the proper anode materials to optimize this technique because the electrolytic products strongly depend on these materials as well as the operating conditions such as the current density and temperature [18]. Various materials have been developed to date as anodes. They are classified as follows: carbon (amorphous carbon, graphite), noble metal or metal oxides (Pt, IrO_2, RuO_2) and non-noble metal oxides (PbO_2, SnO_2, TiO_x). Since early times, electrolysis has been used for the treatment of dye waste water by anodic oxidation of saline to generate sodium hypochlorite solution, which is used for decolouration of dye waste water. Effluent water treatment has also been performed using ozone generated by oxidation of water with PbO_2. In recent years, BDD electrodes have been developed, and they have proved to be effective for detoxifying treatment of waste water containing dye, organic acids, phenols, soluble polymers and so on [19–21]. It has also been shown that diamond-like carbon, is effective for the degradation of persistent organic fluoro compounds [22]. The efficient degradation is attributable to hydroxyl radicals electrogenerated at the BDD anode, resulting in complete degradation to CO_2 [18]. Interestingly, this decomposition mechanism is quite similar to that with TiO_2 photocatalyst. BDD electrodes are also effective in the degradation of organic additives in plating baths, and hence a possibility of recycle use of plating bath is also demonstrated.

It has also been demonstrated that decomposition of harmful organic chlorides like DDT can be achieved by electrolysis using cobalt complex mediator like hydrophobic vitamin B_{12} under photo irradiation [23].

Furthermore, a novel flow and circulating system using a Pd tube as a cathode has been developed for electrocatalytic hydrogenation, and this

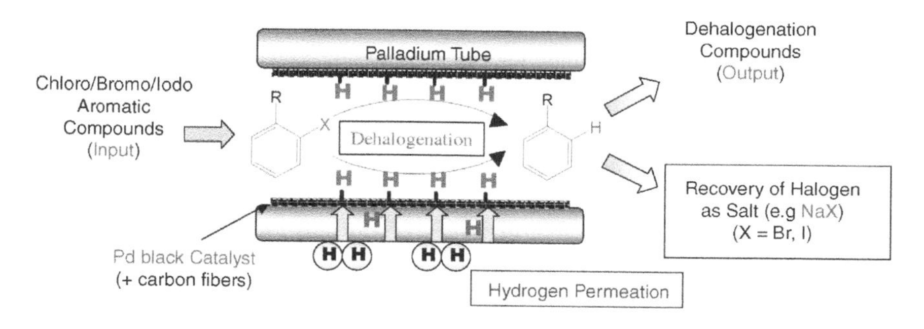

Figure 7.9 Electrocatalytic dehalogenation system

system was shown to be highly effective for the dechlorination of polychloro aromatics, as shown in Figure 7.9 [24,25].

Conversion of chlorofluorocarbon (CFC) to useful substances using gas diffusion electrodes such as Au, Cu, Pd and In has been attempted, as shown in Eq. 7.4.

$$CCl_2F_2 \xrightarrow[\text{1 M aq. NaOH, 7 atom}]{e} CH_4 + CH_2F_2 \quad (7.4)$$

REFERENCES

1. Tang, C.W. and VanSlyke, S.A. (1987) *Appl. Phys. Lett.*, **51**, 913–915.
2. Mitschke, U. and Baeuerle, P. (2000) *J. Mater. Chem.*, **10**, 1471–1507.
3. Tang, C.W. (1986) *Appl. Phys. Lett.*, **48**, 183–185.
4. Günes, S., Neugebauer, H. and Sariciftci, N.S. (2007) *Chem. Rev.*, **107**, 1324–1338.
5. O'Regan, B. and Grätzel, M. (1991) *Nature*, **353**, 737–740.
6. Hagfeldt, A. and Grätzel, M. (2000) *Acc. Chem. Res.*, **33**, 269–277.
7. Heilmeier, G.H. and Zanoni, L.A. (1964) *J. Phys. Chem. Solid*, **25**, 603–611.
8. Yoshida, M., Uemura, S., Hoshino, S., Takada, N., Kodzasa, T. and Kamata, T. (2005) *Jpn. J. Appl. Phys.*, **44**, 3715–3720.
9. Beaujuge, P.M. and Reynolds, J.R. (2010) *Chem. Rev.*, **110**, 268–320.
10. (a) Nonaka, T. and Baizer, M.M. (1983) *Electrochim. Acta*, **28**, 661–665. (b) Baizer, M.M. (1984) *Tetrahedron*, **40**, 935–969. (c) Chum, H.L. and Baizer, M.M. (1985) *The Electrochemistry of Biomass and Derived Materials*, ACS Monograph Series, ACS: Washington D.C.
11. Park, K., Pintauro, P.N., Baizer, M.M. and Nobe, K. (1985) *J. Electrochem. Soc.*, **132**, 1850–1855.
12. Li, W., Nonaka, T. and Chou, T.-C. (1999) *Electrochemistry*, **67**, 4–10.
13. Smith, C.Z., Utley, J.H.P. and Hammond, J.K. (2011) *J. Appl. Electrochem.*, **41**, 363–375.

14. (a) Griesbach, U., Fischer, A., Stecker, F., Botzem, J., Pelzer, R., Emmeluth, M. and Waldvogel, S.R. (2009) Method for electrochemically cleaving lignin on a diamond electrode. WO 2009138368 A1, filed May 11, 2009 and issued Nov. 19, 2009. (b) Griesbach, U., Fischer, A., Stecker, F., Botzem, J., Pelzer, R., Emmeluth, M. and Waldvogel, S.R. (2011) Process for the electrochemical cleavage of lignin at a diamond electrode. US Patent 20110089046 A1, filed May 11, 2009 and issued Apr. 21, 2011. (c) Stecker, F., Fischer, A., Kirste, A., Waldvogel, S.R., Regenbrecht, C., Schmitt, D., (2014) Process for the preparation of vanillin. US Patent 20140034508 A1, filed July 3, 2013 and issued Feb. 6, 2014. (d) Stecker, F., Fischer, A., Kirste, A., Voitl, A., Wong, C.H., Waldvogel, S.R., Regenbrecht, C., Schmitt, D., Hartmer, M.F. (2014) Process for producing vanillin from vanillin-comprising compositions. US Patent 20140046099 A1, filed July 3, 2013 and issued Feb. 13, 2014.
15. Hori, Y. (2008), *Electrochemical CO_2 Reduction on Metal Electrodes* in *Modern Aspects of Electrochemistry*, Vol. 42 (ed. Vayenas, C. G., White, R. E., Gamboa-Aldeco, M. E.), Springer-Verlag, Chapter 3.
16. Nakata, K., Ozaki, T., Terashima, C., Fujishima, A. and Einaga, Y. (2014) *Angew. Chem. Int. Ed.*, **53**, 871–874.
17. Ercoli, R., Silvestri, G., Gambino, S., Guainazzi, M. and Filardo, G. (1973) *Ger. Offen.* 2,235,882, 2 Jan; *Chem. Abstr.*, **78**, 97190h.
18. Panizza, M. and Cerisola, G. (2009) *Chem. Rev.*, **109**, 6541–6569.
19. Foti, G., Gandin, D., Comninellis, C., Perret, A. and Haenni, W. (1999) *Electrochem. Solid-State Lett.*, **2**, 228–230.
20. Gandini, D., Mahé, E., Michaud, P.A., Haenni, W., Perret, A. and Comninellis, Ch. (2000) *J. Applied Electrochem.*, **30**, 1345–1350.
21. Rodrigo, M.A., Michaud, P.A., Duo, I., Panizza, M., Cerisola, G. and Comninellis, Ch. (2001) *J. Electrochem. Soc.*, **148**, D60–D64.
22. (a) Ochiai, T., Iizuka, Y., Nakata, K., Murakami, T., Tryk, D.A., Koide, Y., Morito, Y. and Fujishima, A. (2011) *Ind. Eng. Chem. Res.*, **50**, 10943–10947. (b) Ochiai, T., Iizuka, Y., Nakata, K., Murakami, T., Tryk, D.A., Fujishima, A., Koide, Y. and Morimoto, Y. (2011) *Diamond Relat. Mater.*, **20**, 64–67. (c) Ochiai, T., Moriyama, H., Nakata, K., Murakami, T., Koide, Y., Morito, Y. and Fujishima, A. (2011) *Chem. Lett.*, **40**, 682–683.
23. Shimakoshi, H., Tokunaga, M. and Hisaeda, Y. (2004) *Dalton Trans.*, 878–882.
24. Fuchigami, T. and Tajima, T. (2006) *Electrochemistry*, **74**, 585–589.
25. Kawabata, Y., Naito, Y., Saitoh, T., Kawa, K., Fuchigami, T. and Nishiyama, S. (2014) *Eur. J. Org. Chem.*, 99–104.

8

Examples of Commercialized Organic Electrode Processes

Toshio Fuchigami

Organic electrosynthesis can be performed under mild conditions such as room temperature and normal pressure, and does not require hazardous oxidants or reductants, such as heavy metals. Moreover, it produces less waste compared to ordinary chemical synthesis. In spite of such advantages, commercialized organic electrode processes are limited, in contrast to inorganic industrial processes [1–3].

8.1 AVENUE TO INDUSTRIALIZATION

Organic synthesis usually requires heating. The starting materials and products are easily decomposed to decrease yields and selectivity at high reaction temperatures. In contrast, organic electrosynthesis is a synthetic chemical process that can be operated under mild conditions such as room

Fundamentals and Applications of Organic Electrochemistry: Synthesis, Materials, Devices, First Edition. Toshio Fuchigami, Mahito Atobe and Shinsuke Inagi.

temperature and normal pressure, which avoids heat deterioration of organic compounds. If optimum electrolytic conditions are established, organic electrosynthesis could be a superior synthetic procedure with excellent yield and selectivity. In order to decrease activation energy, various catalysts are commonly employed in chemical reactions, but many catalysts are expensive hence recovery and recycling of catalysts is very important.

In electrolytic reactions, electrodes are electroconductive interfaces where electrons are transferred to and from substrates, as well as acting as the catalyst (electrocatalyst), therefore a costly catalyst is not necessary. Furthermore, the applied potential and current for the electrochemical reaction can be inexpensively and precisely controlled. Thus, organic electrosyntheses have many technical advantages over ordinary organic synthetic processes.

However, organic electrosynthesis also has some disadvantages. Ordinary chemical reactions are homogeneous, while the reaction field of electrolysis is a heterogeneous interface, therefore electrosynthesis has a productive drawback. Moreover, quite differently from inorganic electrode processes like salt electrolysis, in organic electrosynthesis the working electrode providing products is usually either the anode or the cathode. The counter electrode is usually not used for the formation of valuable products except for the paired synthesis developed by BASF, as explained below. Although the main role of the counter electrode is for current flow to the working electrode, the maintenance of the electrode is necessary, resulting in an increase in running cost. When the cathode is a working electrode, the lifetime of the anode as a counter electrode becomes a barrier to industrialization. In general, the anode is readily corrosive and therefore expensive electrode materials have to be used for it. As a result the maintenance cost of the counter anode is often much higher than that of the working cathode. These problems make industrialization of organic electrosynthesis difficult and, as mentioned earlier, it must be recognized that organic electrosynthesis has severe limitations. In spite of this, cathodic hydrocoupling of acrylonitrile has been used for more than 40 years worldwide. This is because the product, adiponitrile, is profitable because of its versatility and in high demand because of its excellent physical properties. The cathodic hydrocoupling method is also superior to other synthetic procedures from an energy cost aspect. Initially, the process was operated using a divided cell, but later electrolysis without a diaphragm was developed in order to decrease the electric power cost of the process.

8.2 EXAMPLES

8.2.1 Electrosynthesis of Adiponitrile

6,6-Nylon is produced by catalytic hydrogenation of adiponitrile to form hexamethylenediamine, followed by dehydration-condensation with adipic acid, as shown in Eq. 8.1. Since 6,6-nylon has excellent strength and durability, it is widely used in tyre cord, synthetic clothes, engineering plastics etc. Hexamethylenediisocyanate as a starting material for polyurethane is also produced from hexamethylenediamine. Because of the benefits of nylon and polyurethane, cathodic hydrocoupling of acrylonitrile is an essential technology for daily life.

$$NC(CH_2)_4CN \xrightarrow{[H]} H_2N(CH_2)_6NH_2$$

$$n\ H_2N(CH_2)_6NH_2 + n\ HOOC(CH_2)_4COOH \xrightarrow[-nH_2O]{} \left[HN(CH_2)_6NHCO(CH_2)_4CO \right]_n \tag{8.1}$$

Cathodic hydrocoupling of acrylonitrile has been known since the 1940s, but the yield of adiponitrile was not high initially. In 1960, Baizer at Monsanto improved both the yield and current efficiency of the hydrocoupling by using quaternary ammonium salts as the supporting electrolyte [2]. In 1963 and 1965, Monsanto developed and commercialized, respectively, the electrohydrodimerization of acrylonitrile, which was produced at low cost on a large scale using the SOHIO method established at that time (Eq. 8.2). An aqueous solution containing a large amount of quaternary ammonium salts (about 40%) was used to dissolve hardly-soluble acrylonitrile at high concentration. Thus, the Monsanto process adopted a homogeneous electrolytic system.

$$2\ CH_2{=}CHCN \xrightarrow[H_2O]{4e} NC(CH_2)_4CN \tag{8.2}$$

On the other hand, the Asahi Chemical Company (now the Asahi Kasei Corporation) in Japan developed emulsion cathodic hydrocoupling of acrylonitrile using the thermomorphicity of acrylonitrile and water containing 10% quaternary ammonium salt, and started operation of the process in 1971 [4]. Both processes were operated in a divided cell equipped with a separator-like cation exchange membrane at the early stage of the processes.

Later on, both companies improved their processes, and in the mid-1980s they established processes with undivided cells in order to reduce the higher cell voltage caused by separator resistance. To develop the undivided cell process, they focused on (i) high selectivity as well as less corrosion, (ii) establishment of electrolytic conditions with stable long-term operation and systems for recovery of quaternary ammonium salt/purification of products, and (iii) development of a facility in which the explosive gas mixture of acrylonitrile and oxygen can be treated safely.

The cathodes used in this process are Hg, Pb or Cd, which have higher hydrogen overpotential because hydrogen generation as a side reaction decreases the current efficiency for the hydrocoupling. On the other hand, Fe, Ni or Pb is used for the anode, and these have lower oxygen overpotentials to prevent the anodic oxidation of organic compounds. The supporting electrolyte is a mixture of quaternary ammonium salts and inorganic salt. Quaternary ammonium salts such as ethyltributylammonium salt are effective not only for increasing the selectivity of adiponitrile formation but also for protection of the cathode from corrosion. Inorganic salts such as potassium phosphate and alkali metal borate increase the conductivity of the electrolyte and prevent corrosion of the anode.

The detailed operation conditions and performances of this coupling process have been reported by several workers [1–4] and an example of operation conditions is as follows:

Cathode: Pb
Anode: Fe (Ni 9%)
Supporting electrolyte: $(EtBu_3N)_2HPO_4 + K_2HPO_4 + K_2B_4O_7$
Current density: $2\,kA\,m^{-2}$
Temperature: 55 °C
Current efficiency: 90.7%

At the present time more than 300,000 tonnes per year of adiponitrile is manufactured by cathodic hydrocoupling of acrylonitrile worldwide, and in 2010 24% of this adiponitrile was manufactured by cathodic hydrocoupling by whole production.

8.2.2 Electrosynthesis of Aromatic Aldehydes

The electrochemical production of acetals of aromatic aldehydes like anisaldehyde and *p*-tolualdehyde has been running for more than 30 years at BASF, as shown in Eq. 8.3 [2]. Oxidation of the methyl group on aromatic rings provides aromatic aldehydes, but ordinary chemical methods using oxidants often cause overoxidation, resulting in the formation of carboxylic acids as a by-product. The electrochemical oxidation of toluene derivatives in methanol provides anodically stable dimethylacetals, which are readily hydrolyzed to form aldehydes and methanol. The recovered methanol can be recycled for anodic acetalization. This process is operated in the capillary gap cell developed by BASF, as shown in Figure 6.26 [5].

$$R\text{-}C_6H_4\text{-}CH_3 \xrightarrow[\text{MeOH}]{-4e,\ -4H^+} R\text{-}C_6H_4\text{-}C(OMe)_2 \longrightarrow R\text{-}C_6H_4\text{-}CHO \qquad (8.3)$$

R = Me, OMe

8.2.3 Paired Electrosynthesis of Phthalide and *t*-Butylbenzaldehyde

Recently, BASF commercialized the paired electrosynthesis of phthalide and *t*-butylbenzaldehyde dimethylacetal, as shown in Eq. 8.4 [2]. The products are useful as plant protectors, additives for plating baths, ultraviolet absorbers, aroma chemicals, fungicides and so on. Phthalide has been prepared by classical catalytic hydrogenation of phthalic acid anhydride, but this method requires pure hydrogen gas and high pressure. On the other hand, during anodic dimethoxylation of *t*-butyltoluene, protons are predominantly reduced at the cathode to generate hydrogen gas. BASF made great efforts to find a compatible reduction process that can be used instead of proton reduction. Eventually, the combination of anodic dimethoxylation of *t*-butyltoluene and cathodic reduction of phthalic acid diester was found to establish the paired electrosynthesis, as shown in Eq. 8.4. Cathodic reduction of phthalic acid dimethylester produces phthalide and methanol, while the resulting methanol is used for anodic dimethoxylation of *t*-butyltoluene to provide dimethylacetal in good yield. In this paired electrosynthesis, the balance of methanol, proton and electron is remarkable.

Cathode

COOMe

COOMe

+4e, +4H$^+$

90%

O + 2MeOH

O

Anode

2MeOH +

−4e, −4H$^+$

90%

CH_3

$CH(OMe)_2$

Total

COOMe

COOMe

+

+4e, +4H$^+$

−4e, −4H$^+$

O +

O

CH_3

$CH(OMe)_2$

Supporting salt: $MeBu_3N^+MeSO_4^-$

(8.4)

This process runs in an undivided capillary cell with a stack of bipolar round graphite electrodes (Figure 6.26) [5]. The electrodes have a centre hole, and are separated by spacers and connected in series. Since quaternary ammonium salt is used as the supporting electrolyte, cathodic reduction of methanol is suppressed and reduction of *o*-phthalic acid dimethyl ester proceeds highly efficiently without formation of by-product. Since efficiency at both electrode reactions is high, the energy efficiency of the paired synthesis is much better than that of conventional processes.

8.2.4 Electrochemical Perfluorination

Electrochemical perfluorination is a process in which all the hydrogen atoms in a starting organic molecule are substituted with fluorine atoms without elementary fluorine generation during electrolysis. J.H. Simons at

3M achieved the electrochemical perfluorination of organic compounds in anhydrous liquid HF using nickel electrodes to provide perfluorinated products for the first time in 1941 [6,7]. He is a pioneer of electrochemical perfluorination and this method is called Simons' process. The process uses an undivided cell at low temperature to keep HF as a liquid (the boiling point of HF is 19.5 °C). Electrochemical perfluorination of organic compounds containing oxygen, nitrogen or sulfur atoms forms salts with anhydrous HF, which provide good conductivity for the anhydrous liquid HF solution. On the other hand, in the case of hydrocarbon, salts such as KF and NaF must be added to impart conductivity and allow the electrochemical perfluorination process. From carboxylic acids (its chlorides), sulfonic acids (its chlorides) and trialkylamines, the perfluorinated products are obtained in good yields, as shown in Eqs. 8.5–8.7 [8]. However, in many cases the yield for electrochemical perfluorination is rather low because of carbon–carbon bond cleavage during electrolysis [8]. The products, perfluoroalkyl carboxylic acids and sulfonic acids, are useful as detergents and lubricants.

$$CH_3SO_2Cl \xrightarrow[\text{Anhydrous HF}]{-ne} CF_3SO_2F \qquad (8.5)$$

$$\underset{(X=Cl,\ F)}{C_7H_{15}COX} \xrightarrow[\text{Anhydrous HF}]{-ne} C_7H_{15}COF \qquad (8.6)$$

$$(C_3H_7)_3N \xrightarrow[\text{Anhydrous HF}]{-ne} (C_3F_7)_3N \qquad (8.7)$$

The reaction mechanism has been discussed for many years [8–11]. One mechanism involves anodically generated fluorine radical as the vital reagent for perfluorination and another involves electrogenerated highly oxidized nickel fluorides such as Ni_2F_5, NiF_3 and NiF_4 on a nickel anode surface. These act as fluorinating reagents and are electrochemically regenerated at the nickel anode. During the electrolysis, partially fluorinated products have polarity and they stay in the electrolyte to be subjected to further electrolysis.

The final perfluorinated products are non-polar and their specific density is very high, therefore they precipitate from liquid HF onto the cell bottom as a liquid.

The electrochemical perfluorination process now runs on a large scale at 3M in the USA and at the Central Glass Company and the Mitsubishi

Materials Electronic Chemicals Company in Japan. Mitsui Chemicals in Japan also produces NF_3 by electrochemical perfluorination of NH_3 [12]. NF_3 is used as an etchant and cleaning gas for apparatus used in the chemical vapour deposition (CVD) technique.

8.2.5 Other Examples

8.2.5.1 3,6-Dichloropicolic Acid

The Daw Chemical Company have commercialized the electrochemical reduction of 3,4,5,6-tetrachloropicolic acid in aqueous solution to 3,4-dichloropicolic acid, which is useful as a precursor to agrochemicals (Eq. 8.8) [13]. As explained in Chapter 4 (Fig. 4.5), the high regioselectivity is attributed to the controlled orientation of the substrate at the cathode surface owing to the dipole moment of the molecule.

$$\xrightarrow[-2Cl^-]{2e,\ 2H^+} \qquad (8.8)$$

8.2.5.2 β-Lactam Derivative

The Otsuka Chemical Company have commercialized the electrosynthesis of the antibiotic substance β-lactam derivative (GCLE), as shown in Eq. 8.9. The electrolysis is operated in a two-phase system of dichloromethane and water, and the reaction proceeds through anodic oxidation of chloride ions.

$$\xrightarrow[\text{aq. NaCl/CH}_2\text{Cl}_2,\ \text{Pt-Pt}]{-e} \quad \xrightarrow[\text{DMF}]{NH_3} \text{(GCLE)} \qquad (8.9)$$

8.2.5.3 Cysteine

Electrochemical reduction of cystin to cysteine is used in many countries, including the USA, Japan, China and some European countries (Eq. 8.10).

Since separation of cysteine from cystin is quite difficult, the electrolysis has to be continued until all the starting cysteine has been completely consumed. This means that at the final stage of the electrolysis electricity is consumed mainly for the reduction of protons to evolve hydrogen gas, resulting in a decrease in current efficiency, which is a problem that has still to be overcome.

$$HOOC\underset{NH_2}{C}HCH_2S{-}SCH_2\underset{NH_2}{C}HCOOH \xrightarrow[\text{Carbon or Ag cathode}]{2e,\ 2H^+} 2\ HSCH_2\underset{NH_2}{C}HCOOH \quad (8.10)$$

8.2.5.4 *Tetramethylammonium Hydroxide*

Electrochemical dialysis of tetramethylammonium chloride uses a cationic ion exchange membrane to produce chlorine-free and highly pure tetramethylammonium hydroxide in Japan (Eq. 8.11). In order to avoid contamination of chlorine in the product, anodically generated chlorine gas is removed completely from the dialysis system.

$$Me_4NCl \xrightarrow[\text{Water}]{-e,\ -1/2\ Cl_2} Me_4NOH \quad (8.11)$$

8.2.5.5 *Other Examples*

BASF manufactures 2,5-dimethoxy-2,5-dihydrofuran and dihydrophthalic acid from furan and *o*-phthalic acid, respectively (Eq. 8.12). They also produce acetoin by anodic oxidation of cyclohexanone with an iodine mediator.

$$\text{furan} \xrightarrow[Br^-/MeOH]{-2e} \text{MeO–(2,5-dihydrofuran)–OMe} \quad (8.12)$$

Furthermore, SNPF in France produces anti-inflammatory phenoprofen by the cathodic reduction of *m*-(α-chloroethyl)phenyl phenylether in the presence of carbon dioxide (Eq. 8.13) [14].

$$\xrightarrow[\text{Mg sacrificial anode}]{2e,\ -Cl^-,\ CO_2} \qquad (8.13)$$

Electrosynthesis of the artificial sweetener maltol and *m*-hydroxybenzyl alcohol used to be commercialized in Japan, but this process had to be stopped because of its high cost compared to alternative chemical processes and the lower price of products made in other countries.

As explained above, organic electrochemical processes often compete with chemical ones therefore the development of electrochemical processes that are superior to chemical ones is necessary. It should be emphasized that organic electrosynthetic processes that address the need for low emissions are highly promising [2].

REFERENCES

1. Macdonald, D.D. and Schmuki, P. (eds) (2007) *Electrochemical Engineering, Encyclopedia of Electrochemistry*, Vol. 5, Wiley-VCH Verlag GmbH.
2. Pütter, H. (2001) *Organic Electrochemistry*, 4th edn (eds H. Lund and O. Hammerich), Marcel Dekker, Chapter 31.
3. Genders, J.D. and Pletcher, D. (1990) *Electrosynthesis. From Laboratory, To Pilot, To Production*, The Electrosynthesis Co. Inc., New York.
4. Shimizu, A. (1998) *Catalysts & Catalysis*, **5**, 269.
5. Beck, F. and Guthke, H. (1969) *Chem.-Ing.-Tech.*, **41**, 943–950.
6. Simons, J.H. (1949) *J. Electrochem. Soc.*, **95**, 47–52.
7. Simons, J.H. (1986) *J. Fluor. Chem.*, **32**, 7–24.
8. Suriyanarayanan, N. and Noel, M. (2008) *J. Solid State Electrochem.*, **12**, 1453–1460.
9. Hollitzer, E. and Satori, P. (1986) *Chem-Ing-Tech.*, **58**, 31–38.
10. Dimitrov, A., Rüdiger, S., Ignatyev, N.V. and Datcenko, S. (1990) *J. Fluor. Chem.*, **50**, 197–205.
11. Sartori, P., Ignat'ev, N., Jünger, C., Jüschke, C. and Rieland, P. (1998) *J. Solid State Electrochem.*, **2**, 110–116.
12. Tasaka, A., Kawagoe, T., Takuwa, A., Yamanaka, M., Tojo, T. and Aritsuka, M. (1998) *J. Electrochem. Soc.*, **145**, 1160–1164.
13. Edamura, F., Kyriyacou, D. and Love, J. (1980) US Patent 4217185; *Chem. Abstr.* (1981), 94, 22193.
14. Chausaard, J., Troupel, M. and Robin, Y. (1984) *J. Appl. Electrochem.*, **19**, 345–348.

Appendix A

Examples of Organic Electrosynthesis

A.1 ELECTROCHEMICAL FLUORINATION

Anodic fluorination and anodic methoxylation of ethyl phenylthioacetate [1,2]

Anodic partial fluorination of organic compounds usually uses a poly (hydrogen fluoride) complex of amine or ammonium salt as the fluorine source and supporting electrolyte. In an undivided cell, hydrogen evolution by cathodic reduction of proton takes place as well as the desired anodic fluorination of a substrate. In an acetonitrile solution, fluorination proceeds via an anodically generated cation intermediate, whereas methoxylation occurs selectively in a methanol solution (Eq. A.1). Optimal conditions for anodic fluorination depend on the substrates used. It is necessary to optimize the combination of reaction media and supporting electrolytes.

Anodic fluorination of ethyl phenylthioacetate proceeds via oxidation of the sulfur atom followed by deprotonation of its α-position. The deprotonation step is promoted by the effect of an electron-withdrawing group substituted at the α-carbon to the phenylthio group. The monofluorinated product at the α-position of the ester group has to be relatively stable if it is to be used as a building block for further transformation.

Fundamentals and Applications of Organic Electrochemistry: Synthesis, Materials, Devices, First Edition. Toshio Fuchigami, Mahito Atobe and Shinsuke Inagi.

PhS–CH₂–CO–OEt (Ethyl phenylthioacetate) → [$-2e, -H^+$; Et_3N-3HF/MeCN; 2 F/mol] → PhS–CHF–CO–OEt
Ethyl phenylthioacetate → [$-2e, -H^+$; Et_3N-3HF/MeOH; 2 F/mol] → PhS–CH(OMe)–CO–OEt (A.1)

Figure A.1 shows a typical electrolytic cell for anodic fluorination, equipped with a platinum plate anode and cathode (2 cm × 2 cm). The cell is filled with an electrolytic solution containing ethyl phenylthioacetate (196 mg, 1 mmol), Et_3N-3HF (1.6 ml, 10 mmol) and acetonitrile (8.4 ml). Constant current electrolysis (20 mA, current density 5 mA cm^{-2}) for 2 h 41 min (2 F mol^{-1}) is conducted with stirring. The solution is then neutralized by saturated $NaHCO_3$ and extracted with ethyl acetate (30 ml × 3). The organic phase is washed with brine (100 ml) and dried over Na_2SO_4. The solvent is removed under reduced pressure and the crude product is purified by silica gel column chromatography to give the desired fluorinated product in 50–70% yield. If methanol is used instead of acetonitrile, the methoxylated product is obtained selectively.

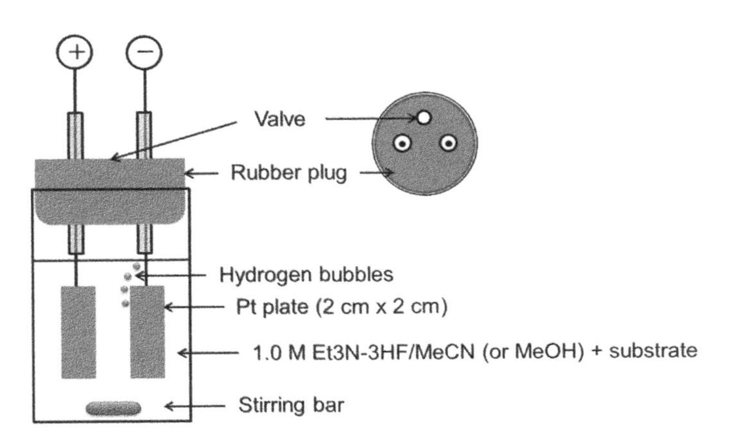

Figure A.1 Illustration of set up for anodic fluorination by the constant current method

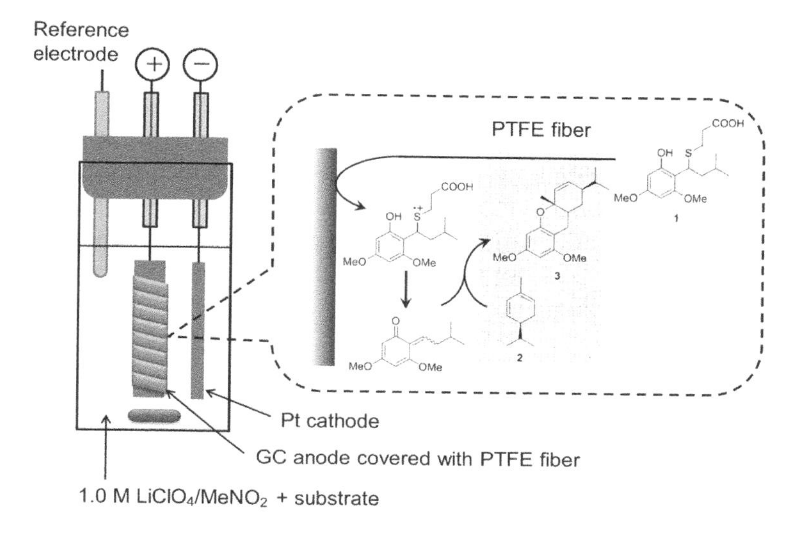

Figure A.2 Illustration of electrolytic set up with a hydrophobic electrode and euglobal synthesis

A.2 ELECTROSYNTHESIS USING A HYDROPHOBIC ELECTRODE

Synthesis of chromane from terpene and electrogenerated quinomethane [3]

A hetero Diels–Alder reaction of quinomethane and terpene is regarded as a biogenetic pathway for euglobal derivatives. Electrochemical generation of quinomethane and subsequent cycloaddition with terpene is possible to obtain chromane, a basic structure of euglobal (Figure A.2).

Sulfide **1** (0.1 mmol) and α-phellandrene **2** (0.4 mmol) are dissolved in 1.0 M $LiClO_4/CH_3NO_2$ (15 ml). Constant potential electrolysis (1.2 V vs. Ag/AgCl) is performed using a glassy carbon anode covered with poly(tetrafluoroethylene) (PTFE) and a platinum plate cathode. After the passage of 1.2 F mol^{-1} of charge, the product is extracted with hexane (10 ml) several times. The organic layer is dried over Na_2SO_4 and product **3** is confirmed by TLC.

A.3 NATURAL PRODUCT SYNTHESIS USING ANODIC OXIDATION

Synthesis of Aniba neolignan by anodic oxidation of phenol derivatives [4]

Although anodic oxidation of phenol derivatives is very useful for obtaining alicyclic compounds from aromatic compounds, its versatile reaction paths make it difficult to produce the desired product selectively. If reaction conditions are optimized, anodic oxidation of phenols can be a powerful tool for obtaining the desired products, which are difficult to prepare by conventional organic synthesis.

According to Eq. A.2, cation **A**, which is generated by anodic oxidation of phenol **1**, undergoes cycloaddition with isosafrole to give a natural product, Aniba neolignan.

$$\mathbf{1} + \mathbf{2} \xrightarrow[\text{LiClO}_4\text{/MeOH-AcOH}]{-2e,\ -H^+} \text{Aniba neolignan } (\mathbf{3}) \qquad \text{(A.2)}$$

Phenol **1** (100 mg), isosafrole **2** (500 mg) and $LiClO_4$ (200 mg) are dissolved in methanol/acetic acid (2:1, 30 ml) and used as the electrolytic solution. In a glassy carbon beaker as anode, constant current electrolysis (10 mA, current density 0.19 mA cm^{-2}) is carried out for 130 min using a platinum wire cathode. The solvent is removed and the product is purified by silica gel column chromatography.

A.4 KOLBE ELECTROLYSIS

Diester synthesis from monoester of adipic acid [5]

The Kolbe reaction, which involves one-electron oxidation of carboxylate and subsequent decarboxylation to form its dimer, is very effective in producing higher alkanes. For example, the monoester of adipic acid can be converted to its dimeric compound (Eq. A.3).

$$2\,CH_3OOC(CH_2)_4COOH \xrightarrow[CH_3ONa/CH_3OH]{-2e,\ -CO_2} CH_3OOC(CH_2)_8COOCH_3 \qquad \text{(A.3)}$$

An electrolytic cell equipped with a platinum anode and a cathode (25 mm × 30 mm, distance 5 mm) is filled with a methanol solution (250 ml) containing adipic acid monomethyl ester (120 g, 0.75 mol), sodium methoxide (4.1 g, 0.075 mol) and pyridine (10 ml, 0.12 mol). Constant current electrolysis (1.1 A, current density 0.13 mA cm^{-2}) is performed for 23 h. During electrolysis under these conditions heat generates to induce reflux of the reaction mixture. After cooling to room temperature, acetic acid (20 ml) is added to the mixture to acidify it. The solvent is removed under reduced pressure and the residual solid is dissolved in ether (500 ml). The filtrate is washed with aqueous $NaHCO_3$ and water, and dried over $CaSO_4$. The product is obtained by distillation under reduced pressure.

A.5 INDIRECT ELECTROSYNTHESIS USING A MEDIATOR

Ketone synthesis by indirect anodic oxidation of alcohol using iodine mediator [6]

The cation species of iodine is generated by anodic oxidation of iodide ions and is used for the oxidation of alcohol to give a ketone product and recovery of iodide ions. The reaction can be mediated by a catalytic amount of iodide without metal oxide (Eq. A.4).

In an electrolytic cell, an aqueous solution (15 ml) of potassium iodide (2.49 g, 0.015 mol) and alcohol (0.06 mol) is produced. Constant current electrolysis using platinum or carbon anode is conducted for 4–15 F mol^{-1} of electricity. The organic layer is separated and the water layer is extracted with diethyl ether. The combined organic phase is washed with aqueous $Na_2S_2O_3$. The product is purified by distillation.

$$R^1R^2CH{-}OH \xrightarrow{KI/H_2O/t\text{-}BuOH} R^1R^2C{=}O \qquad (I^+ \rightarrow I^-;\ -2e,\ \text{Anode}) \qquad (A.4)$$

$R^1 = R^2 =$ [bicyclic structure] 93%

$R^1 = C_6H_{13}$, $R^2 =$ Me: 99%

$R^1 = R^2 =$ Cyclohexyl: 74%

$R^1 =$ Ph, $R^2 =$ Me: 100%

A.6 ELECTROSYNTHESIS OF CONDUCTING POLYMERS

Electropolymerization of pyrrole

Electropolymerization of aromatic compounds proceeds at the electrode surface and a thin film is obtained. These conducting polymers are used for conductive materials, electrochromic devices and catalysts. A typical mechanism of oxidative polymerization of pyrrole is shown in Eq. A.5.

$$\xrightarrow{-e} \qquad \xrightarrow{-2H^+} \qquad \text{(A.5)}$$

In a beaker-type cell, platinum plate electrodes (1 cm × 1 cm) and a saturated calomel electrode (SCE) as a reference are set up. An aqueous

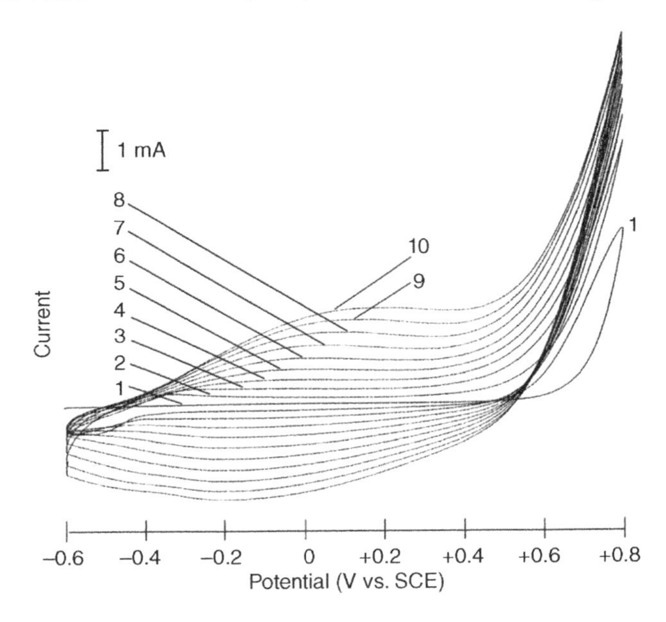

Figure A.3 Cyclic voltammograms of pyrrole during electropolymerization

solution (50 ml) of pyrrole (0.1 M) and $NaClO_4$ (0.1 M) is added to the cell. A potential sweep from −0.6 V vs. SCE to +0.8 V vs. SCE at a scan rate of 100 mV s^{-1} involves polymerization on the working electrode, which is monitored as shown in Figure A.3. The working electrode covered with polypyrrole is purified by washing with water.

REFERENCES

1. Fuchigami, T., Shimojo, M. and Konno, A. (1995) *J. Org. Chem.*, **60**, 3450–3464.
2. Fuchigami, T., Yano, H. and Konno, A. (1991) *J. Org. Chem.*, **56**, 6731–6733.
3. Chiba, K., Arakawa, T. and Tada, M. (1996) *Chem. Commun.*, 1763–1764.
4. Shizuri, Y. and Yamamura, S. (1983) *Tetrahedron Lett.*, **24**, 5011–5012.
5. Haufe, J. and Beck, F. (1970) *Chem. Ing. Tech.*, **42**, 170–175.
6. Shono, T. Matsumura, Y., Hayashi, J. and Mizoguchi, M. (1979) *Tetrahedron Lett.*, **20**, 165–168.

Appendix B
Tables of Physical Data

Table B.1 Potential window of organic solution for electrochemical reactions (Pt working electrode)

Solvent	Supporting electrolyte	Potential (V vs. SCE)	
		Cathodic side	Anodic side
AcOH	AcONa	−1.0	+2.0
Acetone	n-Bu_4NClO_4	−1.0	+1.6
MeCN	$LiClO_4$	−3.0	+2.5
MeCN	Et_4NBF_4	−1.8	+3.2
DMF	n-BuN_4ClO_4	−2.8	+1.6
DMSO	$LiClO_4$	−3.8	+1.3
MeOH	$LiClO_4$	−1.0	+1.3
MeOH	KOH	−1.0	+0.6
CH_2Cl_2	n-BuN_4ClO_4	−1.7	+1.8
THF	$LiClO_4$	−3.2	+1.6
Sulfolane	$NaClO_4$	−4.0	+2.3
$MeNO_2$	$Mg(ClO_4)_2$	−2.6	+2.2
Propylene carbonate	Et_4ClO_4	−1.9	+1.7

Fundamentals and Applications of Organic Electrochemistry: Synthesis, Materials, Devices, First Edition. Toshio Fuchigami, Mahito Atobe and Shinsuke Inagi.

Table B.2 Oxidation potentials of typical organic compounds

Compound	Electrolyte/solvent	Potential (V)	Reference electrode
Aromatic compounds			
Benzene	$NaClO_4$/MeCN	2.00	Ag/Ag^+
Toluene	$NaClO_4$/MeCN	1.93	Ag/Ag^+
Anisole	Pr_4NClO_4/MeCN	1.76	SCE
Biphenyl	$NaClO_4$/MeCN	1.48	Ag/Ag^+
Fluorene	$NaClO_4$/MeCN	1.25	Ag/Ag^+
Naphthalene	$NaClO_4$/MeCN	1.31	SCE
Pyridine	$NaClO_4$/MeCN	2.2	SCE
Thiophene	$NaClO_4$/MeCN	2.10	SCE
Pyrrole	$NaClO_4$/MeCN	0.46	SCE
Olefins			
Ethylene	Bu_4NBF_4/MeCN	2.90	Ag/Ag^+
Cyclohexene	$NaClO_4$/MeCN	1.95	Ag/Ag^+
Styrene	$NaClO_4$/MeCN	1.90	SCE
Nitrogen and sulfur compounds			
Acetamide	Et_4NClO_4/MeCN	2.00	SCE
Aniline	Buffer/H_2O	1.04	SCE
N-Methylaniline	Na_2SO_4/H_2O	0.70	SCE
Nitrobenzene	0.1 M HCl/50% acetone/H_2O	0.58	NHE
Thiophenol	CF_3CO_2H/CH_2Cl_2	1.65	Ag/Ag^+
Dimethyl sulfide	0.1 M HCl/MeOH	0.86	Ag/Ag^+
Diphenyl disulfide	$LiClO_4$/MeCN/CH_2Cl_2	1.75	Ag/Ag^+
Alcohols			
Methanol	Bu_4NBF_4/MeCN	2.73	Fc/Fc^+
Ethanol	Bu_4NBF_4/MeCN	2.61	Fc/Fc^+
Isopropyl alcohol	Bu_4NBF_4/MeCN	2.50	Fc/Fc^+
t-Butyl alconol	Bu_4NBF_4/MeCN	2.60	Fc/Fc^+
Ally alcohol	Bu_4NBF_4/MeCN	2.65	Fc/Fc^+
Benzyl alcohol	$NaClO_4$/MeCN	>2.00	Ag/Ag^+
Phenol	$NaClO_4$/MeCN	1.04	Ag/Ag^+

Table B.3 Reduction potentials of typical organic compounds

Compound	Electrolyte/solvent	Potential (V)	Reference electrode
Halogen compounds			
Chloromethane	Et_4NClO_4/DMF	−2.76	SCE
t-Butyl chloride	Et_4NClO_4/DMF	−2.60	SCE
t-Butyl bromide	Et_4NBr/DMF	−2.19	SCE
t-Butyl iodide	Bu_4NBF_4/DMF	−1.91	SCE
Carbonyl compounds			
Acetone	Et_4NBr/DMF	−2.84	SCE
Acetophenone	LiOH/75% Dioxane	−1.26	SCE
Formaldehyde	pH = 8	−1.22	NHE
Acetaldehyde	pH = 9.1	−1.51	NHE
Benzaldehyde	NH_4Cl/40% EtOH	−1.32	SCE
Quinones			
1,4-Benzoquinone	50% EtOH	0.71	NHE
1,4-Naphtoquinone	50% EtOH	0.49	NHE
9,10-Anthraquinone	95% EtOH	0.16	NHE
Olefins			
Styrene	Bu_4NI/DMF	−2.45	SCE
trans-Stilbene	Bu_4NI/DMF	−2.30	SCE
cis-Stilbene	Bu_4NI/DMF	−2.07	SCE
Aromatic compounds			
Benzene	Bu_4NBr/Me_2NH	−3.42	Ag/Ag^+
Naphthalene	Bu_4NBr/Me_2NH	−2.53	Ag/Ag^+
Anthracene	Bu_4NBr/Me_2NH	−2.04	Ag/Ag^+
Pyrene	Bu_4NBr/Me_2NH	−2.29	Ag/Ag^+
Biphenyl	$NaBPh_4$/THF	−2.68	Ag/Ag^+
Furfural	Britton and Robinson/ H_2O	−1.04	SCE
2,6-Dimethylpyridine	Bu_4NI/DMF	−2.85	Ag/Ag^+
Nicotinamide	Britton and Robinson/ MeOH	−1.34	Ag/Ag^+
Nitrogen and sulfur compounds			
Nitromethane	pH = 7/H_2O	−0.88	SCE
Nitrobenzene	pH = 7/80% dioxane	−0.62	SCE
Diphenyl disulfide	Bu_4NI/DMF	−2.75	Ag/Ag^+
Methyl phenyl sulfone	Bu_4NBr/DMF	−2.41	SCE
Benzenesulfonic acid	Me_4NCl/dioxane	−1.50	SCE

Table B.4 Physical properties of typical solvents

Compound	Molecular weight	Boiling point (°C)	Melting point (°C)	Density d (g/cm^3) (25 °C)	Viscosity η milli-poise (25 °C)	Relative permittivity ε_r (25 °C)	Dipole moment μ ($\times 10^{-30}$ C·m)	Donor number D_N	Acceptor number A_N	Self-ionization constant pK_1
Acetic acid	60.05	117.8	16.64	1.049 2*1	12.2*1	6.15	5.67	—	52.9	14.5
Acetone	58.08	56.2	−95.4	0.784 5	3.02	20.7	9.76	17.0	12.5	—
Acetonitrile (AN)	41.05	81.6	−45.7	0.776 6	3.39	35.95	13.06	14.1	18.9	28.5
Chloroform	119.38	61.27	−63.49	1.489 1*1	5.55*2	4.724	3.40	—	23.1	—
N,N-Dimethylformamide (DMF)	73.10	158	−61	0.944 3	7.96	36.71	12.86	26.6	16.0	—
Dimethyl sulfoxide (DMSO)	78.14	189.0 (Decomp.)	18.55	1.096	19.6	46.6	—	29.8	19.3	~32
Ethanol	46.07	78.32	−114.15	0.785 1	10.78	24.3	5.63	20	37.1	18.9
Hexamethyl-phosphoric triamide (HMPA)	179.20	235	7.20	1.024*3	—	29.6	—	38.8	10.6	—
n-Hexane	86.18	68.7	−94.3	0.659 4*1	3.258*1	1.90	0.00	—	0.0	—
Methanol	32.04	64.75	−97.68	0.786 6	5.42	32.6	5.63	19.0	41.3	16.7
Nitrobenzene (NB)	123.11	210.80	5.76	1.198 6	18.11	34.82	14.03	4.4	14.8	—
Nitromethane (NM)	61.04	101.2	−28.6	1.131 2	6.27	35.94	11.53	2.7	20.5	19.5
Propylene carbonate (PC)	102.09	241	−49	1.19	25.3	64.4	—	15.1	18.3	—
Pyridine (Py)	79.10	115	−41.5	0.977 9	8.824	12.01	7.17	33.1	14.2	—
Tetrahydrofuran (THF)	72.11	65.0	−108.5	0.880	4.6	7.39	5.67	20.0	—	—
Water	18.01	100.0	0.00	0.997 1	8.903	78.54	6.47	18.0	54.8	14.0

*1 20 °C
*2 22.8 °C
*3 30 °C

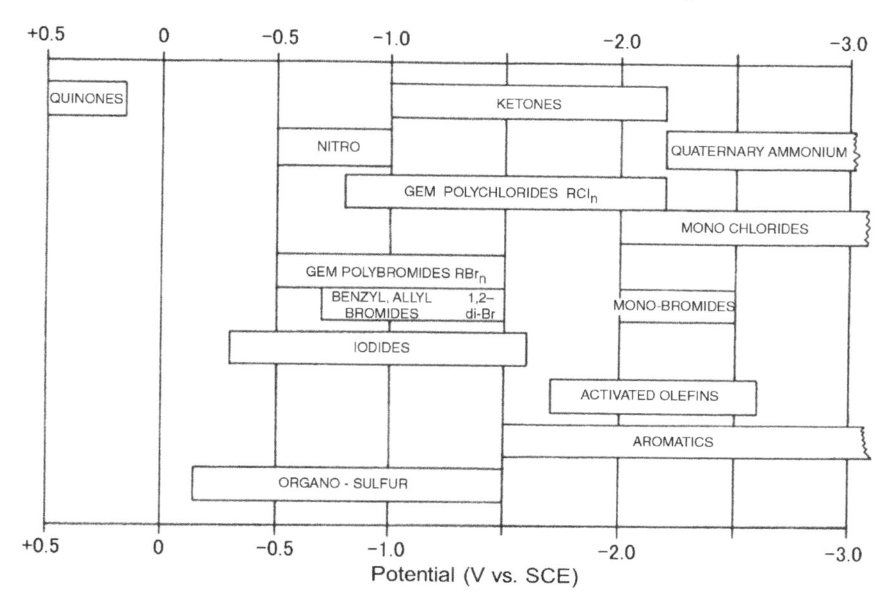

Figure B.1 Electrode potential regions for reduction of functional groups

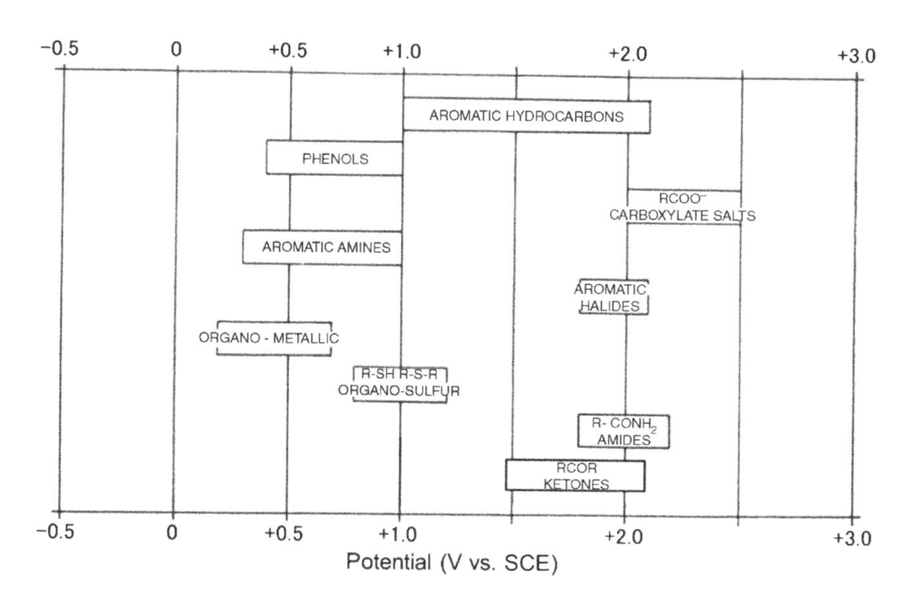

Figure B.2 Electrode potential regions for oxidation of functional groups

Index

Fundamentals and Applications of Organic Electrochemistry: Synthesis, Materials, Devices, First Edition. Toshio Fuchigami, Mahito Atobe and Shinsuke Inagi.

L

M

N

O

P

Q

R

S

Printed and bound by CPI Group (UK) Ltd, Croydon, CR0 4YY

12/01/2025

14624491-0001